Wild
Life

Wild Life

Finding My Purpose in an Untamed World

Dr. Rae Wynn-Grant

GET LIFTED BOOKS

A zando IMPRINT

NEW YORK

GET LIFTED BOOKS

Get Lifted Books is an imprint of Zando
zandoprojects.com

First Edition: April 2024

Cover design by Evan Gaffney
Cover photo © Celeste Sloman / Trunk Archive

Library of Congress Control Number: 2023948634

978-1-63893-040-2 (hardcover)
978-1-63893-041-9 (ebook)

10 9 8 7 6 5 4 3 2 1

Manufactured in the United States of America

To my grandfathers,
who pushed me "straight ahead" into the wild.

To Eleanor Sterling,
who bravely broke the rules and changed my life.

Contents

Prologue

B-b-b-bear!" I stammered as my brain, frozen with fear, struggled to connect with my mouth.

Deep in the hot, arid wilderness of western Nevada, I was face-to-face with a black bear.

In the desert-like brush on the side of a canyon nestled in the Pine Nut Mountains, he stood barely six feet away from me. His dark fur was rendered light brown by the dust covering every inch of his muscular body, which seemed impossibly massive in the light of day. His perked ears twitched, ready to capture every sound I made, and his luminous eyes pierced me in place.

I'd been trying to capture this bear for a week and a half, and thus far he'd eluded me. Days of laying traps and tracking had been futile. But today, as I was distractedly rebaiting the trap with a can of tuna fish, he'd silently crept up on me.

The pristine silence of the mountain air was punctuated by the strong puffs of the bear's exhales. I didn't dare move, every fiber of my being coiled with tension. I searched the recesses of my mind for the training I had received so many times, but nothing surfaced.

Suddenly, he began stomping one of his massive front paws on the ground, shaking his head back and forth and snorting loudly. Our eyes locked.

Even as a fledgling bear biologist, I knew what this meant.

All five hundred pounds of thick fur and solid muscle lunged at me, razor-sharp fangs ready to tear me apart.

"Bear!" I screamed as my brain and voice fully reengaged, and I stumbled backward. Had I been able to conjure my training, I would have remembered exactly what to do: back away slowly and quietly, or make myself small and nonthreatening. Instead, my instinct kicked in, and I did the exact opposite: I turned and ran.

Down into the canyon I sped, kicking up rocks and dirt, creating a dusty cloud that choked me as I gasped for air, my lungs straining at the full-out sprint. I tripped and fell as sagebrush bushes clawed at my body, catching my clothes and tearing at exposed skin. But no wound that rough terrain created compared to what would happen if the bear caught up to me.

Don't look back, don't look back, I repeated over and over in my head, terrified tears streaming down my face. I felt the bear's presence behind me, the earth rumbling as his heavy footsteps hit the ground. His grunts of exertion sounded inches away, raising the hairs on the back of my neck.

While I scrambled through the brush, my mind raced. What had I been thinking, going into the field without my backpack of supplies? Without my bear spray, of all things? As much as I loved being out in the wild, why did I always feel one step behind? And why did it seem like I always had to learn lessons the hard way, making major mistakes that caused me to doubt my own judgment?

After what could have been a second or an eternity, I finally saw my ATV come into view, along with the intern who'd accompanied me into the field and had stayed behind with our supplies while I rebaited the trap.

"Bear!" I screamed again as I neared him, running and stumbling and sobbing hysterically. An ATV had never looked so beautiful.

At the steepest point of the canyon, close to where the incline began to level off, I lost my footing and began rolling down the ravine. My arms and hands flew up to protect my face and head from the dirt and rocks. Unsure if the bear was still chasing me, I scrambled back up the side of the valley, eyes fixed on the ATV and survival.

My battered body reached the vehicle, and I jumped on and fired up the engine. The intern and I sped away from the site. Only as I was heading toward safety did I finally look back.

Heart hammering in my chest, I searched for my pursuer. Through tear-filled eyes, all I could see were plumes of dust settling back to the ground.

Introduction

The closest contact the average American will have with a wildlife ecologist is through a television screen: men traversing our world's jungles, deserts, and other topographic marvels, standing in front of a camera while they interact with wild animals and the environments around them. This, too, was my introduction to the mystery and intrigue of the natural world.

The adventure shows I watched as a child riveted me. I gawked in wonder at David Attenborough, Marlin Perkins, Jim Fowler, Jeff Corwin, and Steve Irwin as they showed me landscapes and species half a world away. The sprawling terrains and unique animals fascinated me, fueling dreams of my future: a wild life of days tracking animals under the crowded canopies of rainforests and of nights sleeping under the starry skies of the savanna. A life of adventure and discovery, far away from cold concrete and steel.

Ecologists like myself are concerned with the beings—human, animal, and otherwise—that exist in community together. We observe the relationships integral to the survival of our very fragile ecosystems. Every living thing in an ecosystem knows the steps to an intricate dance. From the trees that provide shelter and clean air to the fungi that decompose and recycle and the animals that consume and produce energy, all beings are vital to their environment. When one resident misses a step or changes the choreography, all others suffer. Climate change, soil erosion, deforestation, mass extinctions, and pandemic diseases are all examples of these missteps—ones that can result when we don't respect the nonhuman relatives with whom we share space.

As a wildlife ecologist, my job is to rigorously study our world and then develop the best ways to protect it, particularly through helping to manage the fickle relationships between animals and humans. We all exist in a web of interdependence. And unless we can find a way to live together in dissonant harmony, we cannot survive.

Throughout my studies, one thing I didn't anticipate was that the knowledge I gained in the field, studying predators and their prey, would apply to my own life. As a young Black mother and a professional, I've built a career for myself in a space dominated by older white men and charted my own path in a society riddled with ill-fitting expectations. These lessons have inspired my thoughts and activism not only on conservation and climate, but also on race and identity. Animals and ecosystems have a lot to teach us about mutual respect and the importance of caring for one another in mutually beneficial ways.

Observation is a major part of my work as an ecologist, as is recognizing that every location—whether tropical jungles in Africa or bustling streets in American cities—is its own type of wilderness. These stories contain my observations from a career spanning nearly two decades and what I've absorbed from witnessing the shifting relationships between humans, animals, and places. From locations as varied as the cloud forests of Madagascar to the crowded pavement of Manhattan, the ecosystems I've called home have taught me to embrace dissonance, celebrate Earth's diversity, and champion those most affected by the system's failings. My perception of community has flipped from one of tenured knowing to one of growing curiosity. Yet even with a PhD and a career's worth of experience, I am still learning.

To dedicate one's life to studying animals and the spaces they inhabit is to live and grow with them. It requires constant adaptation, self-protective ferocity, and, occasionally, acceptance. May my experiences guide you in making sense of our place in this world, finding your sense of self, and seeking paths to live within and protect our varied ecosystems. And may you be inspired to surrender yourself to a wilder life.

Chapter 1

E ven though I've traveled the world, California has always felt like home to me. I was born in San Francisco and lived briefly in Cincinnati, Ohio, before the Golden State's warm sun beckoned my family to return. Instead of moving back to the more diverse San Francisco, my parents chose San Luis Obispo, a small agricultural town on the central coast. My father knew the town well because he'd lived there while attending California Polytechnic State University (Cal Poly) in the 1970s, and he'd been offered a professorship at his alma mater.

As a Black girl in a predominantly white and Mexican American community, I often stuck out. My dark skin and kinky hair were obvious differences that could be neither concealed nor changed, though I didn't actively notice this until first or second grade. My best friend, Rosanna, would tell me, "I wish I had brown skin too—your brown skin is so cool." And people often commented on my family's hair, in awe of its texture and the styles it could achieve. These kinds of remarks might have seemed innocuous, but they served to make me feel self-conscious and hyperaware of the ways in which my Black family and I stood out.

My home life also seemed vastly different from that of my peers. My dad was an architect and a professor, and my mom was a creative writer. Both of them were college graduates, an accomplishment that allowed them to pursue their passions, and they placed a premium on education for my younger brother, Asa, and me. They weren't super strict with rules, but a couple of nonnegotiables have stuck with me over the years. One was bedtime, which my mother enforced even into our teen years. The second was TV viewing. My parents were strong believers that "TV is bad for you," and I accepted this as fact. It often felt like I was the only kid for whom TV was off-limits. During recess, my friends would excitedly discuss the latest episode of their favorite television series, like *Mighty Morphin Power Rangers* or *Rugrats*. I'd hover idly by with minimal input, though I'd sometimes throw in a one-line interjection I'd rehearsed.

My parents probably wouldn't have ever owned a TV if not for my father's one weakness: professional basketball. Tucked away in our living room closet, the TV was only pulled out when my dad's beloved NBA games were broadcast. He was faithful to two teams: the Golden State Warriors out of hometown loyalty and the Chicago Bulls for sheer fun. Asa and I were allowed to watch the games with him, but we had to leave the room during commercial breaks. My parents were determined to prevent us from becoming obsessed with consumption.

As we got older, Asa and I begged our parents to let us watch more TV. They eventually said yes, but they weren't going to let us rot our brains by watching the latest popular shows—we were only allowed to watch educational programs.

Right away, I fell in love with nature shows. For hours I'd sit there, entranced by the brave adventurers and wild animals co-existing in the world's mythic locales. I had an affinity for large carnivores like tigers and lions, and rainforests seemed like the most interesting places on earth, lush with a vast array of plant and animal life. Much of the science presented was beyond my understanding, but that did nothing to dampen my interest. The hosts' passion kindled something deep within me, and I longed to have adventures in these places that seemed beyond my reach.

One night at the dining room table, around age seven, I announced that when I grew up, I would be a nature-show host who lived in the jungle and told people about animals. My parents said, "Okay," and continued eating. They were neither supportive nor critical, a kind and passive way of letting me know these discussions were welcome.

Every time I watched a nature show, I consumed messages that highlighted the dangers our planet faced. Rainforests were being destroyed and oceans were being polluted. Bald eagles and other animals were almost extinct due to the use of pesticides. Nature shows not only showed me the natural world's beauty, but also warned of its degradation at the hands of irresponsible human behavior. I felt an almost instinctual urge to do something. Too young to take monumental action, I did what I could: picking up litter along my walk to and from school each day, crushing aluminum cans before recycling, and informing anyone who would listen about the plight of blue whales.

At our core, I think all human beings have an innate desire to be stewards of a healthy, balanced planet, whether that desire manifests in a commitment to picking up trash around our neighborhood or being directly involved in research or shaping policy. In my child's mind, hosting a nature show equated to being out there—boots on the ground and "doing something" for our planet, showing the world the magnitude of what was going on. Like the shows I watched did for me, I could then get people excited about nature, warn them about all the horrible things happening to our planet, and encourage them to take action.

Even as my passion for the environment grew, I was subconsciously convinced my dream was unattainable. At the time, the only hosts I saw on TV were white men, many with British or Australian accents. I don't recall the hosts ever explaining how they'd landed their jobs or talking about wildlife ecology as a career—where to go to college, what to study, how to grow into the role. In the absence of a tangible road map, I concocted my own list of what I believed the qualifications were: white, male, middle-aged, and foreign. I could love nature, be inspired by it and learn from it, but I began to accept that there was no pathway for me to become a nature-show host. Regardless, I never stopped watching the shows and loving the vicarious adventures they took me on.

These nature shows were my portal into the wilderness. My family was urban, all of us born and mostly raised in big cities. It was a comfortable environment for us, and we delighted in exploring the diverse cultures, sites, and intricacies to be experienced. We didn't have enough money for vacations, so when we traveled, we often visited family in other cities or accompanied my dad to conferences. When we did visit the great outdoors, it was only for the occasional trip to the beach.

My urban upbringing was peaceful and love-filled, even though my experiences in nature, as I understood it from the shows I watched, were few.

WHAT I CONSIDER my first real wilderness experience was a class trip. A yearly trip for the sixth-grade class, the students waited in anticipation for the day we'd finally get the chance to wave goodbye to our parents, board the school bus, and take the four-hour drive to world-renowned Yosemite National Park.

When I stepped off the bus, the air in Yosemite felt completely different, cleaner and more alive. I inhaled deeply through my nose and into my lungs. The sunlight, less piercing and intense than at the beaches I was used to, added a golden glow to my brown skin. I had experienced many different cities, towns, and beaches in California but had never seen anything like the gorgeous forests that cover the Sierra Nevada mountain range where Yosemite lies. The towering trees, combined with the majestic land formations, blew me away. Growing up with an architect father, I'd been taught to appreciate the beauty in human design, the careful craft of the man-made environment. Yosemite gave me a newfound respect for the architecture nature has provided for us.

Our first day began with a hike through the lower-elevation areas of the park, where we could view some of Yosemite's most iconic features, like Half Dome and Vernal Fall. During the trek, I ate an entire apple, including the core, per the instruction of our guide, whose advice was to leave no trace of our presence as we explored the pine forests. I felt like a rugged, outdoorsy survivalist, prepared to tackle the elements if our group got lost in the woods.

The second night—which happened to be my twelfth birthday—brought snow, which was my first experience with California snowfall. The dark clouds blanketing the sky signaled that the weather would move through swiftly, while we stayed cozy inside the large cabin that housed my entire class. As we settled in for an evening of card games and truth or dare, we began to see light flakes that melted upon hitting the ground. Although it was nowhere close to being a blizzard, to a young urbanite from the coast, it was a birthday miracle.

My classmates and I giddily ran outside, unconcerned about getting our pajama pants and lightweight sweatshirts wet. We were only able to catch a few snowflakes on our eyelashes and tongues before the lodge staff whisked us back inside. Thinking the adults were merely trying to restore order after the snow-induced hysteria, we were shocked to learn that real danger was afoot: a black bear had been spotted wandering around our campground.

We scrambled to the windows, hoping to catch a glimpse of the bear. Away from city lights, the campground outside was pitch black, and the snow, combined with twenty sixth graders clamoring around a tiny window, made it impossible to make out much of anything. But we all pretended that we saw the bear—hooting and hollering, squealing with delight and screaming with fake fear. Black bears are one of the biggest and, arguably, the most iconic land mammals in the entire state—and this was the first wild animal I had ever "seen" in its natural habitat, save the cute sea urchins and hermit crabs in the tide pools at the beach. Even I swore up and down that I'd spotted the animal, and the feeling was exhilarating.

WHEN I WAS in middle school, my family uprooted from California and moved to Norfolk, Virginia. My dad had been teaching at Cal Poly for several years, and though he enjoyed his work, the position was unstable, and he longed for new teaching experiences, different challenges, and upward mobility. He was thrilled to secure a job at Hampton University, which is an HBCU (historically Black college or university). As one of very few Black architects at the time, my dad felt that a large part of his purpose was to inspire young Black students to pursue architecture as a career path.

My parents also wanted to ensure that Asa and I were educated in a better public school system than the ones available to us in California, which were notably terrible in the 1990s. They also wanted us to grow up in a community with a higher Black population, which San Luis Obispo sorely lacked.

Moving from California to the South came with a seismic culture shock. We emigrated from a white neighborhood to one that was almost evenly split between white and Black residents. While I was happy to be around more people who looked like me, I also began to witness firsthand the kind of anti-Black racism that I'd only read and heard about.

As we drove through Norfolk, roads were flanked by Confederate idols—houses with flags and street corners with memorials. There was little apparent wealth in Norfolk, and the economic disparity fell along color lines; white families were generally more well-off than Black families. My family's middle-class, professional positioning upended this convenient binary. California money could stretch a lot further in Virginia, and we moved into a bigger house than our previous home, in a neighborhood with landscaped yards and people who were comfortable in their finances. We'd been transported into an unfamiliar habitat, one with a social history that bled into the present, and we were tasked with the tall order of not just adapting but thriving.

Our neighborhood was squarely middle-class, with homes and yards many would envy, but for two preteens, the most exciting feature stood at the corner of our street: *a hot dog stand.*

This hot dog stand, which we could barely glimpse from our home's second-floor window, became the object of our childhood obsessions. It felt so East Coast, completely foreign from the life we'd known in the Bay. That first week, we endlessly badgered our parents to take us to the hot dog stand until they finally relented. We walked the half block to the stand, which was similar to a New York City–style cart, with a red-and-white-striped umbrella and a hot dog in a bun painted on the front of the cart. Miraculously, there was no line. But when we approached the counter to order, the middle-aged white man running the stand ignored us. He refused to make eye contact with any of us and blatantly ignored my father's attempts to get his attention. We might as well have been invisible.

Shortly thereafter, some white patrons came to the cart and were immediately served. My family and I stood uncomfortably to the side, watching as the white families received their hot dogs, along with smiles and friendly greetings.

My father loudly commented, "This man doesn't want to serve us," and he approached the cart, demanding to know why we weren't being served. We were out of earshot, so I didn't hear the exchange. I could tell from their angry looks and sharp hand movements that the tension was escalating. The white patrons stood on the sidelines, doing nothing but silently witnessing our humiliation.

Clearly, we wouldn't be getting hot dogs that day.

I didn't fully understand what was going on, nor could I articulate why we weren't being served. My dad had shared stories with us about being mistreated in California because he was a Black man. While I'd experienced what I'd now refer to as microaggressions, I hadn't directly experienced being excluded solely based on my race.

In Norfolk, we now had to learn to navigate life in a community of people who saw us—and, even worse, judged us—based on the color of our skin in a very overt manner. Even in a town surrounded by more Black folks than ever before, I was made painfully aware that people are sometimes treated differently based on fundamental aspects of their humanity. This is my earliest recollection of facing what it means to be Black in America. And sadly, that standoff at the corner hot dog stand was far from the last racially motivated conflict I'd witness or endure.

ALTHOUGH ASA AND I are close in age, the gap was just enough that there were a couple of years when we attended different schools. That happened our first year in Norfolk, with him in fifth grade and me in seventh grade. The split occurred again when he was in middle school and I was in high school. Those years we'd make the morning walk to school alone, a departure from the daily companionship we'd had attending the same school in California.

One day during my sophomore year of high school, I arrived home earlier than Asa. As I prepared an after-school snack, I was startled to hear a loud knock on the front door, which I could see from the kitchen. While I hesitated, my mom came down from her second-floor bedroom to answer the door.

There on the threshold was Asa, with terror in his eyes, and a stern-faced, white male police officer standing next to him.

"Ma'am, is this your son?" the officer asked, as my mother reached for Asa and pulled him into the house, clutching him to her body. I watched from the kitchen, frozen. Was Asa in trouble? Had something terrible happened? He didn't look physically hurt, but he seemed scared.

"Yes, this is my son. What's going on?" my mother asked in a brave and confident tone as she protectively pushed Asa behind her.

"I found him walking slowly down the street, ma'am, and I needed to make sure he got home safely—that's all." The officer then launched into a lecture about how my brother could have presented as less "suspicious."

"He's home now—we're all finished here," my mother interrupted calmly as she started to close the door and flashed her hand as a wave goodbye to the officer.

My mother whipped around to face Asa, who began to speak, his voice wavering then and dissolving into a light cry. He'd been walking home from Blair Middle School, like he did every day, backpack bulging with books from all of his advanced classes, causing him to move slowly. The trip home took him from a commercial area of Norfolk into our almost entirely white neighborhood, called Ghent. It was a middle-class neighborhood adjacent to West Ghent, the distinctly wealthier area.

As he walked, a police car started to trail him, and the officers inside eventually asked him to stop. They asked him what he was doing but didn't accept his reply: walking home from school. They then asked where he lived and rejected that answer as well. Asa looked like any other twelve-year-old Black boy in Norfolk, which to the officers meant low-income and a high risk for crime. They suggested he was walking through Ghent because he planned to commit a crime—perhaps he already had, which explained his full backpack.

Listening to Asa, my gut knotted and tears fell from my eyes. Our parents had given us "the talk" more than once. They'd told us stories about my father, merely trying to live life as a young Black professor on majority-white college campuses, frequently being harassed by security. My dad would come home from work furious that he was forced to show his ID everywhere he went at his place of employment. And that was California, where racism was allegedly less overt.

Now here I was, listening to my little brother—the one I felt I was born to protect—tell his own story of racial profiling and police harassment, with equal parts humiliation and fury apparent on his face.

He explained that he'd kept insisting he was walking home. The officer said, "Oh yeah? Well, then, what's your address?" Asa gave the address, and the officer, still not believing that this Black boy could be who he said he was and could be doing what he said he was doing, said that Asa needed to prove that's where he lived. The officer put him in the back of the police car and drove to our house. We can only guess what would have happened if the knock at the door had drawn a stranger instead of my mom.

IN NORFOLK, the confluence of people from different races and economic backgrounds made for a dynamic, and occasionally hostile, public school experience, exemplified by one particular foil, my high school math teacher. Each school day he roared into the school parking lot atop a motorcycle adorned with Confederate-flag paraphernalia, his bald head and outdated handlebar mustache a visible indicator of his intimidating demeanor and regressive mindset. He'd stride into the classroom still wearing his helmet, only to unceremoniously place it in its designated spot on his desk, Confederate flag sticker front and center.

He routinely ridiculed the Black students. Whenever a group of Black boys would make any kind of noise in the hallway during class, he'd say, "The natives are restless." He'd also insinuate that the Black kids in the hallway were mentally inferior, which was why they weren't in our classroom—they were incapable of learning precalculus.

None of us ever felt comfortable speaking to the man, let alone asking him questions about trigonometric identities or polynomial functions. Like many of my classmates, I didn't feel safe in that classroom.

Without proper support and encouragement, my grades began slipping. In my mind, if you had aptitude in a subject, you'd ace your tests in that class. If you failed, it was because you weren't skilled in that area. I'd internalized this message from my father, who spent his days working with students trying to find their path. In his view, not performing well in a subject area—and demonstrating skill in a different area—indicated the route a student "should" take. I compared myself to Asa, who got straight As his entire high school career. No subject was too hard for him, and while he did have to study, school came easily to him. Asa ended up graduating as valedictorian of his class, with the highest GPA in the entire state of Virginia.

Although I wanted to be "smart," I developed a malignant belief that I wasn't good at academics. Not showing aptitude in my math and science classes led me to believe my dream of studying wildlife ecology was unattainable. So I turned to music.

I'd been taking piano lessons since I was five years old, and I was relatively talented. Back in San Luis Obispo, I'd perform in piano recitals, and my teacher helped me learn to play the harpsichord. By the time I was nine years old, I was proficient enough that I toured with the New World Baroque Chamber Music Orchestra.

Even so, music was never my passion—it was a hobby. I didn't choose to get involved in music. My parents made both Asa and me play an instrument (he was a violinist and also talented), and we spent years developing our skills.

But music became less of a hobby and more of an active pursuit when I began participating in a specialized music program after my sophomore year. My skills were adequate, so diving into classical music and applying for a competitive program wasn't particularly intimidating. I had yet to discover the full extent of my talent and potential, and I never could have predicted how much the experience—from the friendships to the mentorship and the discipline required—would impact the rest of my life. I was strengthened in ways that helped me far beyond my performances onstage.

"THIS ISN'T QUITE what we're looking for."

Although Alan Fischer, the chair of the vocal department at the Governor's School for the Arts, was speaking to me, he was making eye contact with his coleader, Robert Brown, who sat at the piano accompanying me. Their not-so-subtle body language communicated what I feared: my hopes were doomed.

I was auditioning for a one-of-a-kind conservatory-style music program, which also happened to be free, as part of Norfolk's public school system. (Picture the movie or TV show *Fame*, and you've got the idea.) It served high school students, but I was a rising eleventh grader and thus attempting to enter in the middle of the standard four-year program. Which meant I had to prove that my voice was on par with the other eleventh graders, who'd already had two years of classical vocal training.

"Let's try this," Mr. Fischer said, slightly exasperated as he tilted his chin toward the ceiling. "You know when you watch cartoons and they'll have a character singing opera—she might be wearing a Viking helmet or something?"

"Yes," I responded, as several classic Bugs Bunny cartoons flashed through my mind.

"Try singing your piece the way that character would: exaggerate the 'opera' sound like they do in the cartoons."

These two men were making a last-ditch effort to salvage my audition. While my light, airy, sweet voice could carry a tune, it wasn't strong enough for this program.

I decided to abandon any self-consciousness and get silly with it. What did I have to lose?

I took a deep breath, lifted my shoulders, and spread my hands out wide. I then projected a long "ahh" note with as much vibrato in my voice as I could muster. It was loud, dramatic, and fun.

"*Yaaaaaasssssssss!*" Mr. Brown shouted, jumping to his feet, away from the piano bench. He stretched his long arm toward my face as if to pull more sound from my body. "*Yaaasssss*, this is it! Give me another note!"

I couldn't believe this was what they'd been trying to elicit from me, but their furrowed brows and their eyes locked onto mine made me realize how serious they were.

I took another deep breath as Mr. Fischer, now also yelling, told me to try a high note with an "eee" sound. A vibrato-heavy "eee" came from somewhere in my diaphragm.

Prior to my audition for the Governor's School for the Arts, my experience as a vocalist had been limited to singing as a run-of-the-mill high school student in chorus class. My primary instrument had been piano, not my voice. Yet there I was, securing my place in a competitive, rigorous program that trained professional singers—some of whom had gone on to become stars in the opera world.

Being accepted into this program and its exclusive community instilled in me tremendous confidence. In public high school, I was someone who dwelled on the periphery—on the outskirts of the circles of both the Black kids and the white kids.

Though I could relate to my peers in either circle, I existed in a sort of social limbo. And at a time when boys were interested in girls with curves, I was lanky, flat-chested, and booty-less. In general, I was shy and insecure, and I preferred to fade into the background.

Of course, that kind of reticence wasn't an option in the Governor's School program. We not only had to perform constantly in front of the faculty and one another, but also were critiqued. Mr. Brown and Mr. Fischer always told us that the critiques weren't intended to demean or discourage us; they were meant to push us to be better. These two men sincerely believed in every single student and often reminded us of their faith in our potential.

Mr. Brown was a force. His six-foot-three frame and loud, resonant voice dominated any space he occupied. He also happened to be a self-taught musical genius who could play numerous instruments, conduct, arrange music, and teach. His larger-than-life personality, power, control, and expertise overflowed onto the students, fertilizing the seed of talent within each one of us.

Mr. Fischer was as opposite of Mr. Brown as he could be: a short Jewish man from New York and a classically trained vocalist. Back in the 1960s, he'd performed in the children's choir at the Metropolitan Opera.

This unlikely duo leveraged their unique strengths to educate and train a bunch of ragtag, awkward teenagers. Nowadays, I think of them in terms of some of the great music partnerships, like Rodgers and Hammerstein or Disney collaborators Howard Ashman and Alan Menken. Individually, they were fabulous; united, they were unstoppable.

Our teachers took their work and their students seriously, which really boosted our self-esteem. It's said that if you set the bar high, then people will rise to the occasion. My classmates and I did just that. Many of us, including me, scored internships or spent senior year with opera companies. I think you'd be hard-pressed to find a single alum from the program who wouldn't say that the inner strength they possess today was forged in those intense years, even if they didn't recognize what was happening at the time.

I became one of the leading voices and was chosen as a soloist by the end of my time at the Governor's School. Both Mr. Brown and Mr. Fischer encouraged me to apply for a conservatory to further my vocal instruction. I loved what I was learning and studying, and I especially loved performing. Yet those childhood memories of nature shows were seared in my memory, and my passion for music couldn't dislodge my desire to pursue a science career.

My parents joined Mr. Brown and Mr. Fischer in imploring me to pursue instruction at a conservatory. Scientific fields were not even part of the conversation, because I'd been earning Cs and Ds in my science and math classes.

Even with poor grades, I still romanticized and dreamed of careers that would allow me to make discoveries about the natural world. But I couldn't deny that I wasn't performing in academics at the same caliber I was performing in voice or piano.

Another factor influencing my parents' advice was finances. We were lower-middle-class, and my parents didn't have a savings account or significant equity in our home. So regardless of whether I attended conservatory or college, I'd probably have to take out student loans. It made sense for me to go somewhere I could land a scholarship—followed by a career where I'd succeed and thus have an income to repay my loans. My persistence in studying biology in college was complicated, scary, and stressful for my parents. They'd always supported my dreams, but this moment was extremely challenging for them.

In December of my senior year, I was accepted to conservatory, and my path seemed laid out before me. Despite my likely success in music, I couldn't shake my passion for science. I may have been a talented classical musician, but my plans felt inauthentic.

I never wanted to disappoint the people who loved and supported me most, but as the months passed and the college applications were due, I had to be more adamant that I wanted a college degree, not conservatory training.

"But don't you love music? You're so good at it!" both of my grandmothers would question me.

"Yes, but I don't think I want to be an opera singer when I grow up," I'd answer. "It's competitive, and even if you are one of the best, that doesn't necessarily mean you're going to have a lot of opportunities."

What I was unable to vocalize was what my heart was telling me: that I needed to continue forward into the unknown, down the path that would lead to faraway adventures in the wilderness, and work to save endangered species from extinction. As far as I knew at the time, there wasn't a school for that, but it called to me nonetheless.

I could tell my family and teachers were concerned that all of my immersion into classical music might have been in vain. "What are you going to do, sing to the animals?" Mr. Fischer asked me at one point. I didn't want to combine the two fields; I wanted to love and appreciate music from a place of deep knowledge and experience, not from a place of labor.

"I'm going to college. I never thought saying that would cause such a problem, but that's my final decision," I said.

My weary parents eventually capitulated and accepted my decision to go to college and pursue my true passion. For perhaps the first time in my life, I decided to trust my instinct and bet hard on myself.

The following fall, I packed my bags and moved to Atlanta to attend Emory University. After being rejected by my top-choice schools, Emory was the highest-ranking school that accepted me, along with offering me a generous financial aid package. Emory labeled itself as an "affirmative action university," and I imagined that institutional priority—combined with my résumé, personal statement, and my good-but-not-great grade point average and SAT scores—allowed the admissions team to look at the entirety of my identity and background when making their decision.

The personal-statement prompt on the application had directed: "Write a personal statement that demonstrates something unique about you." I wrote about the Highway 41 fire, which occurred in 1994 while we lived in San Luis Obispo. I described being a nine-year-old kid, standing outside my house and seeing the ten-foot-high flames on the hills surrounding our town. Ashes fell from the sky like rain, so we used umbrellas when we walked outside. We'd stocked our car with luggage and emergency supplies, ready to evacuate at any moment. The ecosystem we lived within was burning, and all we could do was watch and wait. Living in Virginia had been different—water, rather than fire, did the most damage there. Hurricanes had us sheltering in place, and forests were cut down instead of burned.

I wrote that story, not my journey through music or the racism that drove me out of Norfolk, because that experience had shown me how fragile our environment was, and how our world needed diverse, experienced, trained minds to figure out how to save it.

I was excited about living in Atlanta, learning at Emory, and starting this new, exciting chapter. Leaving the family and friends I loved in Norfolk was hard, and it was a curious tension to feel so supported and cared for at home and in my music community, yet recognize that I didn't belong in Norfolk—nor did I want to. The palpable inequality and racism there had almost broken me at times, and I was determined to set my life on a course of unbridled success, some twisted effort to show Norfolk what it had lost in driving me away.

This mindset might have been unhealthy, yet it was one Asa and I shared. We vowed to never make Norfolk our home. And neither of us ever did.

Chapter 2

With eager anticipation, in the fall of 2003, I arrived in Atlanta, the Black mecca of the American South. With a robust history of Black achievement in education, science, art, business, politics, and activism, Atlanta was a welcomed change from the backwater bays of Norfolk.

Unlike my high school in Norfolk, Emory University made a concerted effort to ensure its students of color felt welcome and safe on campus. The school set up an orientation specifically for minority students, which took place a week before the full-fledged freshman orientation.

I walked into a cavernous lecture hall where the orientation was being held and took a seat. A professor stood at the front of the room and explained that each one of us would be paired with a mentor who was either a junior or senior. Our mentor would walk us through the process of creating a five-year plan, which would help us map out our four years at Emory and one year beyond. Emory's medical school is exceptional, so most of the students in my program were premed. Though I was interested in some of the subject matter that was part of the premed curriculum, such as biology and anatomy, becoming a doctor never appealed to me.

When the professor gave us the go-ahead, I opened my packet and prepared to find my mentor, a dark-skinned and lanky, yet handsome, upperclassman.

"Hi—I'm Ryan." He introduced himself with a bright smile.

Although this was my first time meeting Ryan one-on-one, I had encountered him on move-in day. I saw him ordering people around, walkie-talkie and clipboard in hand, giving out room assignments. He was a sort of "big man on campus," extremely popular among not only the Black students but also the entire campus community. He'd noticed me, too, and we'd even engaged in some mild flirting.

I was happy to be paired with him because he was both kind and extroverted. And I knew his honesty and great sense of humor would make the somewhat serious and daunting task of crafting a five-year plan less stressful.

We sat down together, and he wrote "Rae's Life" at the top of my yellow legal pad. I was supposed to document my hopes and plans for my future, yet my mind was as blank as the page in front of me. But as Ryan and I talked and laughed together, he put me at ease. He was a great listener, and our comfortable rapport helped me open up.

"I really love science, but my grades in those classes always sucked. I do know that I don't want to go to medical school," I told him. "I'm also really good at music, and I enjoy singing. I'm interested in pop culture, too, and journalism. Not that I want to do any of those things, though—music or journalism. I'm just thinking about things I'm good at. . ."

"What do *you* want?" Ryan gently asked me, giving me a special kind of attention that made my heart flutter.

"The only thing I've ever truly wanted to do is host a nature show," I confessed. Due to my consistently poor grades in STEM (science, technology, engineering, and mathematics) classes, it had been years since I'd considered science a viable career option. That conscious reality was compounded by my subconscious belief that ecologists were white men. Or at least the ones on television were. But now, with Ryan's encouragement, for the first time in almost a decade, I let myself dream.

"Write it down," he instructed me.

On the legal pad, underneath "Rae's Life," I wrote, "Host a nature show."

Beneath that goal, I created a time line that included the classes I'd need to take to graduate, the first job I wanted after college, and any aspirations for future degrees. We figured that nature-show hosts were knowledgeable about all things related to the natural world. We weren't able to list all of the concrete steps I'd need to take to reach my goal, but I had to start somewhere, so between the two of us, we decided that it made the most sense for me to major in environmental science.

The five-year plan was helpful, and it felt good to share my dream with someone, but I didn't necessarily walk away from the meeting feeling more confident or secure.

I had a lot of intrinsic motivation to succeed, yet the extrinsic motivators were powerful as well. My dad was practically going broke because of the loans he'd taken out to pay for my higher education, and I wanted more than anything to prove that his investment in me was well-placed. I'd work my butt off to excel in my classes and make my parents proud.

After my heart-to-heart with Ryan, his mild interest in me blossomed into something more. I was certainly attracted to him, and I think part of the reason he was attracted to me was because my talents and career aspirations were so different from many other students, and I was comfortable dreaming this rare dream that was off the beaten path. While my program wasn't the most diverse, the university's large student body consisted of people from all walks of life, with stories radically different from my own. At Emory, I began to understand the concept of community, surrounding myself with people who unequivocally believed in my success.

About a month after the minority-student orientation, Ryan and I began dating. The relationship ramped up slowly, because Ryan respected that I was seventeen and fairly inexperienced in relationships, both emotional and physical. He also knew that I was tackling the wide world of college with extreme focus. Our time together was fun and intense. Being attached to the big man on campus afforded me, a measly freshman, access to the world of upperclassmen. Most of them lived off campus and threw wild parties every weekend, with abundant alcohol, drugs, and sex.

Exposure to this lifestyle made me realize how sheltered I'd been throughout my life. I'd had a curfew throughout high school, and I'd never engaged in underage drinking or even seen any drugs. I'd been the type of teenager who was interested in sex and being seen as "sexy" by guys my age, but the alcohol-fueled, casual sex culture intimidated me. With Ryan by my side, I was protected from any unwanted attention, yet I was also free to try new experiences. I felt so cool, being able to code-switch from being a straight-edge, academic science major during the week to attending parties with one of the most popular guys at school during the weekends. I could dip in and out of both worlds, although I knew the party scene wasn't meant for me. At least not yet. On some level, Ryan's preference for high-risk spaces and that corresponding lifestyle concerned me. But in those early days, I was just enjoying the ride.

MY FIRST SEMESTER, I had a full course load. In pursuit of my environmental science degree, challenges were inevitable. One of the first hurdles I encountered was the math requirement. I hardly slept the night before my first calculus class, my nerves and excitement swirling inside of me when I realized I was finally beginning my science education. When I arrived at the lecture hall, I felt transported back in time to my Norfolk precalculus classroom, steeling myself against harassment from the mustached teacher and surrounded by Confederate memorabilia.

As I slouched into an empty seat, I thought, *I'm probably going to fail this class, like I've failed all of my other math classes over the years.* Thankfully, my negative self-talk didn't have the last word, and I said to myself, *Look, you're in college now—if you want to succeed, you need to take some initiative.*

After that first lecture, I approached my professor at the front of the room. He was the polar opposite of my high school precalculus teacher, in both appearance and demeanor. He was fairly short, dark-haired, chubby, and young. A neatly trimmed beard and mustache distinguished his kind face.

"Hi, I'm Rae," I began. "I'm taking this class because I want to be an environmental science major. But I'm really bad at math."

"No, you're not. No one is bad at math," he replied matter-of-factly.

This response threw me into a spiral of self-doubt, devastating me further as I began to wonder if everyone was good at math except for me.

The professor must have seen the insecurity and disappointment on my face, because he kindly smiled and assured me that everything would be okay. He then suggested that I complete my homework during his office hours, so he could answer my questions as they arose.

The first time I walked into his office, I once again had flashbacks to interacting with my high school precalculus teacher.

I'm sure my body language conveyed my nervousness, and my professor strove to set my mind at ease.

I worked through the equations on my own and got almost every question wrong. The professor then spent about an hour and a half walking me through every single equation, reteaching me the material from class. By the time we finished, I wouldn't say that I understood the material any better, but I did feel that with his help, I could succeed in the class.

One of the most important paradigm shifts I experienced was that it was acceptable to ask for help. In the past, I'd believed that you had to do everything on your own. That somehow, receiving any kind of help diminished your accomplishment. But this professor helped me see that getting a good grade in math could happen many different ways, whether from a student's own labors; from receiving assistance from peers, tutors, or professors; or from a combination of the two. As long as you weren't cheating, it didn't matter how you mastered the material.

And my professor was right—I wasn't bad at math. I ended up getting a C in the class, mainly because of my test grades. When I spoke to my professor about the C, he was unperturbed. He told me he was glad that I had mastered some of the concepts and, more important, that I'd renewed my interest in math. Though I would've loved a better grade, I was proud of myself for gaining confidence in myself and my abilities.

I had low self-confidence rooted in years of being neglected in the public school system. The willingness of this professor to invest in his students and foster their potential had a major impact on my fledgling career as a scientist. For the first time, my interests and academic performance aligned, and my dream no longer seemed impossible.

*   *   *

LIFE AT EMORY was shaping up even better than I'd anticipated. I was gaining ever-increasing confidence in my academic
abilities, I had a boyfriend who was both popular and attentive,
and I was developing close relationships with a diverse group
of girlfriends.

At the beginning of the year, I had a temporary roommate for
the minority orientation program, a young Black woman named
Dana, who was fun and down-to-earth. She was from Oakland,
and we were both excited to meet someone else from the Bay
Area. Unlike me, she'd gone to private school, but her family
wasn't wealthy; she'd attended on scholarship.

That first day on campus, Dana and I were hanging out in
our room, unpacking some of our toiletries. While we talked
and laughed, sharing stories with one another, my birth control
pills fell out of my bag. Our eyes met. I was slightly ashamed and
embarrassed. I'd been taught that things like birth control were
private matters. Not only that, but I'd just met Dana and had no
idea what her views were.

"Is that the pill?" she asked.

"Yeah . . . I started taking it this summer," I confessed.

"Hey, me too!"

Now at ease with one another, our conversation shifted to
boys.

"I actually lost my virginity last night," I admitted.

"Really?"

"I started dating a guy at the end of senior year, and I talked
him into having sex with me before I left to come here."

What I didn't tell Dana was that I'd had a master plan to lose
my virginity. I hadn't wanted to go to college without knowing
at least a little about sex. I wanted to explore, to be free to have
whatever experience I wanted. And for whatever reason, holding

on to my virginity wasn't important to me. I was glad not to have waited.

My friendship with Dana multiplied into close friendships with many other Black girls—especially after I joined the step team. It was an Emory tradition for Black freshmen to join the step team, a rhythmic art form that was well established at HBCUs, in Black fraternity/sorority culture, and in the Black community as a whole. The girls and boys had separate teams, and we performed twice a year. Though Emory wasn't necessarily known for its step team, Dana and I threw ourselves into learning the art, excited to experience something new.

Other than my exploits with Ryan, step team was my primary source of fun. We practiced a lot—usually in the campus parking garages, to ensure that our routines remained top secret. On Mondays and Wednesdays after practice, we'd all go to the late-night diner on campus and stuff ourselves with pancakes. Around 1:00 a.m., I'd sneak into my dorm room and collapse into bed, only to awaken the next day and rush off to calculus class.

Because I was still young, my relationship with Ryan was pretty innocent. We dated for months before we were intimate. One day, we crossed that line, and I shared the news with my roommate, Tina, who had squealed with excitement.

"Do you love him?" she asked me.

I hadn't been prepared for that question. Ryan and I had said "I love you" at the end of our first month of dating.

"I think I love him. I mean, I love him, obviously, as a person, and I like him a whole lot romantically. But I don't know that I'm going to marry him or anything. I don't know. I told him I love him, and I think I'm going to get there really soon," I told Tina.

I did feel close to Ryan. In many ways, he offered a sense of security that allowed me to feel rooted, comfortable, and cared for during the intense transition from high school to college. But the reality was that our relationship was bordering on codependency, something I'd never experienced. As our time together progressed, I started to feel like I was losing, rather than gaining, independence.

WHEN I REFLECT on some pivotal times in my personal and professional growth, I recognize certain people in my community who showed up and offered subtle but powerful ideas and perspectives that helped loosen the shackles on my mind. One of these angels was Dr. Frances Smith Foster.

Dr. Foster was an English professor who specialized in Black American literature and women's studies at Emory. She was a Black woman in her late fifties with short natural hair that was mostly gray. I'd seen her around campus several times, always carrying an armful of books. She often stood out to me, partially because of the confidence she exuded and the way she carried herself when she walked. She moved slowly, as if Emory were built around her. The school had few Black professors, which highlighted Dr. Foster's visibility even further.

All incoming first-year students at Emory University were required to take a freshman seminar. These intimate classes were discussion-based and allowed students to delve deep into a topic of interest. The extensive list of seminar options excited me. I could learn about diverse subjects ranging from art history to neuroscience, and I debated between the wildlife-focused classes. We had to select our top three favorites and rank them in order of preference. My number one choice was primate behavior.

My second choice was a seminar about the science of the brain, which I felt aligned well with my environmental science major. And my third choice was a music class. I hadn't come to Emory to study music, but it remained in my heart and soul. Also, I figured it was highly unlikely I'd be given my last choice, so I may as well put it on my list.

I didn't get placed in any of my top three seminars. I was placed in a class called Becoming a Woman, which was taught by Dr. Foster. It was decidedly not about primates or brains or music. I'd glanced at it during my initial read-through of the seminar offerings. I thought that the title sounded cool, but why would I need to take such a class? I was already a woman.

I begrudgingly walked into Dr. Foster's seminar on the first day and was struck by how different the classroom looked from all of my other courses. Most of my classes took place in massive lecture halls with stadium-style seating. But this cozy room tucked in the English department only contained about ten desks arranged in a circle. Five of those desks were occupied, all by young women of color.

Another Black girl I knew, Brittany, was there. We were familiar with each other and both happy to see a face we recognized. We sat next to each other and learned that neither of us had intended to be in this seminar. And we both hoped that the add/drop period would allow us to select a different option.

Shortly thereafter, Dr. Foster entered the room and sat down in the circle.

"Welcome, ladies." She spoke in a low, resonant voice, in the way I imagine Maya Angelou or Nina Simone would have greeted a classroom.

"Becoming a Woman is a class I designed specifically for you," she said. "It's a literature class for freshman girls of color, where we'll read coming-of-age novels with lead characters who are young women of color." I couldn't recall having ever been assigned a coming-of-age novel about a girl, rather than the classic boy-centered ones like *The Catcher in the Rye* or *Lord of the Flies*. Let alone a novel featuring a girl of color whose racial identity dramatically adds to her life experience.

Suddenly, this class was everything I didn't know I needed.

I pored over the syllabus, perusing the list of books I'd never heard of before but now couldn't wait to devour. Dr. Foster told us that our only homework would be to read the assigned novels and provide thoughtful contributions to our class discussions. She aimed to have a no-pressure seminar that would serve as a space for us to reflect on the ways the stories seeped into our souls—and hopefully elicit the emotional depths that were hiding below the surface of our practiced confidence and competence.

The books we read included *The Joy Luck Club* by Amy Tan and *The House on Mango Street* by Sandra Cisneros, among other tales that rooted us in the transformations happening within our bodies and minds. We saw many nuances of ourselves reflected on the pages of these novels. The characters, whose stories were often set in decades past, seemed young and naive to me, as though they were several development steps behind me. Yet if that was the case, why was I riveted by these stories, gripped by the tension and the angst in the writing?

During the semester my classmates and I spent with Dr. Foster, she not only taught us about literature but also mothered us in many ways—and not always sweetly. One week several students skipped class due to illness, and the rest of us arrived to class with cold symptoms. That day, she gave her only lecture of the semester. She scolded us for not taking good enough care of our bodies and insisted that we do better.

"I know you girls aren't exercising regularly because I go to the gym on campus, and I don't see any of you there!" she said. In response to our blank faces, she continued, "You all do know the connection between exercise and your immune system, right?" We didn't, but we sure learned that day. Dr. Foster wasn't satisfied with us being regular, slightly irresponsible college freshmen. She cared about our well-being as a parent would, and she pushed us to center health and wellness.

"Your health is your biggest ticket to success," she said, ending her lecture and dismissing us from class early, so we could return to our dorm rooms to hydrate and sleep.

Dr. Foster's class was a safe, nurturing space, and it provided tremendous comfort in an otherwise tumultuous freshman experience. The class became a trust circle for vulnerable freshman girls of color. Eventually, I was able to admit that I, much like the protagonists in the novels we read, was a young and naive girl facing womanhood a bit abruptly. As much as I desired independence and enjoyed the buzz that attention from men gave me, perhaps the jump from sheltered child in my parents' home to a young adult in Atlanta was jarring in ways I hadn't expected. Perhaps a slower-paced journey toward maturity would be more comfortable for me.

In mid-October, the students dispersed for fall break, a four-day weekend. My parents flew me home to Norfolk, which surprised me because money was tight and that type of expense wasn't typical. However, my excitement soon turned into a tsunami of emotions I'd spend years working through.

One night, my family and I sat around the dinner table, and my mother announced: "Your father and I are separating." My dad silently hung his head, avoiding eye contact with any of us.

I leave for college, come home a month later, and my whole family falls apart, I thought. In my family, confrontation was rare and uncomfortable, so I'd subconsciously learned to people-please rather than confront. In this situation, that meant filling my mind with toxic words rather than speaking them aloud.

In the wild, animals react with a fight-or-flight response when confronted with fear or danger. People are no different. I fought, not with fists, but with denial.

"What do you mean? You guys have the best relationship! You literally never fight!" I said, beginning with a quiet tone that reached near-shouting levels. "How is this happening all of a sudden?"

I looked over at Asa, who sat stoically while making eye contact with everyone. His posture was more upright than usual, and it seemed like he was also choosing to fight this news. I was angry because I was confused. I'd always felt grateful that my parents got along so well and had such a cool, nontraditional relationship. They had always been a team that took the path less traveled together.

Six weeks after my mother gave birth to me, my parents packed us up and boarded a plane to Fiji. They weren't going for vacation but to participate in social justice protests in support of local communities. As I grew up, living in a household where the television was typically unplugged and pushed into the corner meant that most evenings were filled with stimulating conversations, bonding activities, and memorable fun. In addition to being a writer, my mother freelanced as a conflict-resolution facilitator. Her training, combined with my father's routine mindfulness practices, fostered a relationship devoid of yelling, aggression, or stubbornness.

How can two peaceful people who love each other want to end their marriage? I wondered. *What could be missing after more than twenty years?*

"Nothing happened—this is nobody's fault," my mom said. "We've been working with counselors and therapists, and we believe it's best for us to separate. I'm going to live nearby in Hampton, and your dad will stay here in this house with Asa."

I was crying now. Sobbing, loudly. My dad had tears in his eyes, and my mother's face looked upset but not tearful. She was holding back pain. Looking at both of them, I believed what they were saying. Their words omitted much of what would help Asa and me understand the decision, but their faces demonstrated that they were both devastated to be sharing this news with us.

"Come on, Asa," I said. We both stood up from the table, walked out the front door of our house, and climbed into our parents' car, me in the driver's seat. We drove around Norfolk aimlessly, venting our anger to each other. How could they get to this point? Divorces were for families with dysfunction! How could they tell us this news without giving a reason? Why did this separation have to occur right now, when, for the first time, Asa and I weren't in the same place, to face this challenge together?

Despite having a close, open relationship with our mom and dad, we didn't have a family culture of children challenging parents. Thus, Asa and I didn't feel comfortable demanding more information about what led to their breakup. In the car that day, we decided to double down on our original intention, which we'd been saying to one another since we'd arrived in Norfolk years before: we'd become hyperindependent and escape Norfolk as quickly and permanently as possible. What was the use in staying in a town we found problematic if we didn't even have the comfort and stability of our parents' loving marriage?

I told Asa that, as had been the case throughout our entire lives together as best friends and siblings, I would always be someone he could depend on, and I would do anything for him. At sixteen and eighteen years old, we would strategize our remaining teenage years together and help each other through anything, especially our parents' divorce.

We returned to our family home, now our father's house, calmer yet emotionally exhausted. I called Ryan, who had stayed at Emory during the fall break, and told him the news. I then went to bed without dinner. The next day, my parents and Asa somberly drove me to the airport to catch my flight. I hugged them all goodbye, unsure what our next visit would be like.

After returning to Emory, the first day back in Dr. Foster's class, she asked, "Would anyone like to share something interesting that happened during the break?" I raised my hand and without being called on blurted out, "My parents broke up." I paused to say more but began to cry. My head dropped down on my desk, and I released my pent-up tears. Within moments I felt Dr. Foster's warm, loving hand on my shoulder and rubbing my back.

Whether I was ready or not, becoming a woman was fully in process.

TWO YEARS INTO MY STUDIES at Emory, my parents had officially separated and were living apart. Visits home were unfamiliar, and the pain was palpable. As their marriage ended, I couldn't help but think about how it began. Once upon a time, they were two college kids whom fate unexpectedly brought together across an ocean.

My parents attended college in the 1970s; Mom went to Clark University in Worcester, Massachusetts, and Dad went to California Polytechnic State University in San Luis Obispo.

The same semester, they signed up to study in Copenhagen, Denmark.

One day, my mom and dad were walking toward one another on the street. Black people were rare in Copenhagen at the time, so they both felt compelled to say hi. They detected each other's American accents and stopped to have a conversation. They clicked instantly and began hanging out regularly. Then they started dating and fell in love.

After their respective programs ended, they returned to the United States to finish their degrees, but they kept in touch. Postgraduation, my mother decided to move to California to be with my father.

Though I had no aspirations of meeting my future spouse through a study-abroad program, my parents had always encouraged participation in one because they felt it opened the world to you in new and different ways. I'd even written "study abroad" in the five-year plan I'd written with Ryan's help.

One reason I initially put off studying abroad was that Asa was navigating our parents' separation solo. He'd completely immersed himself in his high school studies, partially because he'd always been a gifted, brilliant student and partially out of survival. He found it easier to stay busy with school and extracurricular activities than to sit in the stillness of our parents' dissolving relationship. By the time I was finishing my sophomore year at Emory, Asa was deciding which of the top ten schools in the country he wanted to attend and was preparing to graduate as his class's valedictorian. With him on a secure track toward independence, I set my sights on my next adventure.

However, I had no idea where I wanted to go. Lots of places seemed attractive—Madrid or Paris. Or perhaps I'd retrace my parents' footsteps in Copenhagen. The Caribbean was also an option because I'd always been drawn to tropical locales.

Though I didn't know where I wanted to spend my semester abroad, there was one thing I did know for sure: after being an environmental science major for two years, I was getting bored. I wanted to learn about the environment, but learning about forests, rivers, wild animals, and landscapes from inside a classroom or from the pages of a textbook was unfulfilling. A few times I had even flipped through the course catalog, wondering if I should change my major. But none of the other offerings felt like a solution. I decided that for the first time in my life, at almost twenty years old, I needed to get into nature and see some wild animals.

During the second semester of my sophomore year, I visited the study-abroad office on campus and met with a counselor to explain what I was looking for.

"I'd like to do the most hard-core wildlife ecology study-abroad program there is," I informed the counselor, who looked at me with confusion.

"What do you mean, 'wildlife ecology'?" he asked.

"I'm an environmental science major, and I want to do a wildlife program—somewhere in the wilderness with wild animals."

As he began curating options, one thing became evident: most of the programs Emory offered were expensive and well beyond my budget. I needed to find a program that cost either the same amount as my Emory scholarship or a bit less.

"Even if it's not focused on wildlife, what's the cheapest program you have?" I asked the counselor.

"Off the top of my head, I don't know," he said, and he told me that he'd do some research and get back to me.

Eventually, he found an opportunity offered by the School for Field Studies, which operated several affordable programs around the world that focused on conservation and environmental science.

I narrowed down my options to two possible locations, Australia and Kenya. Thankfully, I had an aunt who'd studied abroad in Kenya, who told me that while her experience had many ups and downs, it changed her life. So she encouraged me to give it a try.

The program was also based in wildlife management. I'd get to see elephants and zebras—and lions, my favorite large carnivores—all of which I'd only ever seen through a television screen. At the time, it was as close as I could get to being a nature-show host, immersed in the wild.

I applied for the program and wrote my personal statement, explaining that this type of opportunity had been a dream of mine for most of my life. I told my family and all of my friends about my plan, and everyone was excited for me and hoped I'd be accepted. That is, everyone except Ryan.

Ryan and I had gone through a difficult breakup at the beginning of my sophomore year. His high-risk, irresponsible party lifestyle had started catching up to him, causing him to struggle mentally, emotionally, and academically. The biggest problem was that his popularity was at an all-time high. His natural charisma, charm, and genuine warmth toward people allowed him to appear stable, if not thriving, on the surface. Not only this, but his brilliance was always apparent—to his professors, peers, and family, and certainly to himself. But I was hyperaware of, and very concerned about, what was occurring behind the scenes. I could tell that his academic struggles made him feel ashamed, which in turn led to more self-destructive behavior.

Over the course of our brief relationship, it was hard for me to watch his downward spiral, and worrying about his wellness was a constant preoccupation. However, I knew I couldn't take on the responsibility of saving him from himself. Additionally, I'd worked too hard and had come too far to let anyone, especially a love interest, keep me from soaring to the heights I could achieve. In spite of the difficulty of separating myself from him, my academic performance continually improved, and I was pursuing professional goals that spoke to my heart's deepest desire.

Even though we had broken up, in many ways we still acted like we were in a relationship. We were still being intimate, and we still talked all the time. We also remained emotionally entangled, albeit in an unhealthy, codependent way. I didn't feel like I had the freedom to live my own life, to date whomever I wanted and move on from him. Yet at the same time, I didn't feel strong enough to demand these things. My inner monologue of affirmations didn't align with what I was doing in practice. His feelings meant a lot to me, and I was afraid of hurting him. I was worried that if I left our relationship fully, he'd punish me with complete isolation. He was popular and close with my entire social group, and if he decided to ice me out and others followed suit, that would have a huge negative impact on my social life at Emory.

He'd never point-blank told me that he wouldn't be okay if I left the country, but he often hinted at it. And when I told him about my intention to study in Kenya, he wasn't having any of that.

"You can study abroad, but not in the bush!" he said, his face stern. "How am I going to visit you if you don't even have an address?"

"This study-abroad program isn't meant to have visitors. In fact, I think there are rules that you can't have friends or family visit. It's a closed campus, er . . . campsite," I told him.

Ryan seemed a bit panicked, and he started listing other couples on campus where one of the partners had studied in a big city or in the Caribbean, where they could more easily visit each other and keep their relationship strong.

At one point, he insinuated that if I left, I'd lose his love. Conflict erupted within me. I didn't want to lose either Ryan's love or my dream. It was true that his love and attention made me feel protected and secure. Yet I knew our modus operandi was unhealthy. I needed to grow, to go away and have adventures, and to take another step into independence. This was a big opportunity for me to have a grand adventure, and I wasn't going to let him stand in my way.

I was accepted into the program, and I spent the summer between my sophomore and junior years with my mom, scouring different shops and thrift stores for a sturdy backpack and hiking boots.

I was an environmental science major who'd spent minimal time in nature. All my exposure to that world had been via TV and textbooks. Although my family traveled, we always visited urban destinations. We'd never gone camping, and I had zero experience with hiking, pitching a tent, cooking over a fire, or anything else that could be remotely construed as roughing it.

In my mind, I could see young Rae's eyes reflecting the television screen, bright and vibrant with all of the life that could be found beyond the confines of her day-to-day existence. Now, for the first time, I'd be immersed in that world in the flesh.

Chapter 3

As we prepared to land at the Nairobi airport, seeing the sprawling land below reminded me of one of my motivations for applying to this program: a need to visit the continent. I didn't know where my African ancestors had originated, but I imagined they were from somewhere in West Africa. But even spending time in East Africa would allow me to connect with my ancestral homeland for the first time. I felt that stepping foot on the African continent would ground and guide me in a significant and profound way.

Before I knew it, the plane touched down. The program staff met us at the airport, holding a sign that said "School for Field Studies," a beacon in the sea of unfamiliar faces to draw all of us American students. I walked over to the staff member holding the sign, a tall Kenyan man in his early thirties who was dressed in a button-down short-sleeve white shirt and cargo shorts. He waved his hand at me, looking over my head and past me at the white American students behind me.

"Excuse me—I'm waiting for students from an American study-abroad program," he said to me.

"Yeah, yeah—I know," I said.

"This isn't for Kenyans. This is for American students," he asserted.

I was the opposite of what they expected, which has been a theme in my professional life. I looked behind me at the other students as they began making a formation around the staff member. We had all met each other at the gate at JFK Airport, and although we hadn't developed friendships yet, we were able to recognize those who belonged to our small group. I knew I was at the right place, and I knew why he was confused.

"I'm Rae Wynn-Grant and I am American and I'm part of the study-abroad program," I told the staff member.

He was super apologetic. He didn't realize that a Black student was coming, and he, along with many of the staff from that program, didn't have a lot of experience with the Black American identity.

We exited the airport and climbed into a Jeep that was waiting to transport us to our field camp. As we drove down the road, I saw my first wild animal, a marabou stork. I'd spent a lot of time reading textbooks about African wildlife, but I'd never come across the marabou stork. I would later learn that these birds can be five feet or more in height and have a twelve-foot wingspan. When you peer out onto the savanna and see them strutting along and bobbing their heads, they can easily be mistaken for something prehistoric. This first sighting of a majestic animal in its natural habitat solidified in my heart that I had made the right decision—not only to come to Kenya, but also to pursue a career as a wildlife conservation professional.

Our program was located in a camp outside of a big national park in southern Kenya and within a Maasai community. In this part of Kenya, the Maasai lived a deliberately traditional tribal lifestyle, the same one their ancestors had lived for thousands of years.

This was 2005, before cell phones and the internet were readily available in East Africa. So we lived off the grid. We depended on monthly mail runs to Nairobi, which was six hours away, to receive correspondence from home.

About a week into the semester, everything was still new. In a moment of familiarity, each of our professors, all Black Kenyan men with PhDs, walked us through the syllabus. The study-abroad program would be intensive. We'd spend a normal amount of time in class, but we'd also be in the field a lot, going on game drives where we'd spend the entire day out there with the wildlife in the bush.

Since we were screen-free, most of our downtime was spent in conversation. We hung out with the staff and got to know them well. Most of them were pretty young; I think the majority were in their twenties. Some were from the Maasai communities, and they served as a bridge between a traditional lifestyle and a Western one. I came to realize that I had a lot of knowledge and information, specifically about the Black American experience, that the staff wanted to talk to me about.

I was the only Black student in my program. Not only that, I was the only Black student who had *ever* been part of the program. Ironically, it seemed that my identity as a Black woman might prove to be more important in Africa than it had been at home in the US.

"EWW! THERE'S A PUBIC HAIR in my food!" This exclamation, followed by an exaggerated gagging sound, tore through the laughter and conversation of our group. Our attention swiveled to Brian, one of the few male students in our study-abroad program.

Brian pushed back from the table and pointed at a wiry black hair resting atop the avocado slices arranged next to the main dish, beans and rice. Everyone was grossed out and started inspecting their own food for a rogue pubic hair. Some people likely lost their appetite altogether at the mere thought of such a thing embedded in their food.

In the background, the Kenyan staff members were cleaning up the kitchen after cooking for our group. When the commotion erupted, they started darting their eyes at each other. What would the consequences of this grievance be?

I didn't even have to walk over to Brian's table and look at his plate to know what was happening. And I hated it.

"Brian!" I yelled from where I sat, at the table next to his. "Brian, we're in Africa. Black people have kinky hair. That's a hair from someone's head."

An awkward silence filled the air for a couple of minutes while Brian and the rest of the students processed this fact. Brian was embarrassed, as he began to realize that his reaction wasn't merely an innocent mistake. What he'd said was racialized, and the power differential between a white American college student and a Kenyan staff cook from the local village was significant. A complaint about unsanitary food carried weight, even in our field camp's open-air dining hut in the middle of the East African bush. While a stray hair from someone's head falling onto an avocado was neither sanitary nor appetizing, it could certainly be tolerated.

Although we were early on in our Kenyan adventure, this wasn't the first sign that race—and Blackness in particular— would be a major theme during this trip. Confrontations with white privilege and my lack thereof had begun the moment our plane landed. And here I was again, sharing a new perspective with my white classmates—not only about their lack of knowledge of Black hair texture but also about their impulse to ring the alarm rather than extend the benefit of the doubt. Their hasty reactions could cause harm, and they needed to be more aware of that.

During the course of our time in Kenya, I engaged in many conversations with both my fellow students and the Maasai tribe about numerous topics related to race: Black identity, the slave trade, racial injustice, and much more. My white classmates were ready to hear and receive difficult truths, and the Black Kenyan camp staff, community members, and some of my professors wanted to hear reflections from me—a young woman who represented a different part of the African diaspora.

The staff members' knowledge and awareness of the Black American experience varied greatly. I remember having two very different conversations with two different staff members. One of them thought that there were no Black people in the United States. So they were curious, wondering how someone could be Black but fully American, without an African heritage they were able to trace. A different staff member expressed pride, saying it was wonderful that I was young but the first person in my family to be free and not a slave. That statement blew me away. It occurred to me that they had no idea about American history, particularly the role of slavery, the Black experience in the United States, and the time line of events.

It wasn't surprising that they didn't have deep knowledge of Black American history. For one, Kenya wasn't really involved in the Atlantic slave trade. And when I think about how little I learned about African history in school, I imagine they similarly didn't have a lot of reason to learn about American history either.

To help everyone understand and to stop the spread of misinformation, I decided to sit down and teach a full-on history lesson.

Not everybody knew that slavery had happened in the United States. A couple of staff members genuinely thought that any Black people in America had parents who had deliberately immigrated there. I corrected that misconception right away. At the same time, I acknowledged that there were, in fact, many Black folks in the United States with parents who had deliberately immigrated there from Africa.

I was fairly unprepared to tackle these conversations. I sometimes felt as lost in the facts, history, and hope as the staff members who were asking me to teach them. I was challenged with explaining all of this while emphasizing that the 150 years after slavery included more brutality, racism, and violence.

But I also asserted that it was possible for Black people to have a great life in America, and that the Black community was thriving and resilient, and doing our best. "We're all quite safe," I said to a roomful of staff members, who had become my friends, their eyes wide with interest and suspense. "You should come to the United States sometime if you can. I think you'll love it."

Early in the trip, a Maasai community member from the nearby village came up to me and my friends when we were there on a visit. We'd met him before; he had limited English skills, but he'd always made an effort to talk to us. On this day, he looked furious as he approached and started yelling. Before I could even understand what he was saying, I was frightened and taken aback when he pointed his finger directly at me.

"You said everything's okay in America! You said things are fine, but all the Black people—they're dead in the water," he shouted.

Confused, I felt certain that a language barrier was to blame. But he kept asserting that "all the Black people in America are dead in the water."

"That isn't true," I firmly told him. "I don't know what news you're getting or where you're getting it from, but it's impossible for what you're saying to be true—that all the Black people in America right now are dead in some water." I was trying to make sense of it and explained that maybe he was thinking about when Black people had died during the Middle Passage.

"No, no—today they're dead in the water," he insisted.

In confusion and denial, I brushed off his statement as misinformed.

"Everything's fine," I assured him, and I shut down the conversation.

The incident freaked me out, but I also let it go. This wasn't the first language-barrier-induced cross-cultural misunderstanding I'd had.

I didn't think of the confrontation again until weeks later, when one of the staff members returned from the monthly Nairobi mail run. My parents had sent me an oversize manila envelope. It was the first mail I'd received, and I eagerly tore it open. The parcel contained several individual envelopes, a couple of postcards, a CD, and a special edition of *Time* magazine about Hurricane Katrina. The cover displayed a stark image of floating bodies, and the feature article detailed that over one thousand people had died, almost all of them Black residents of New Orleans. I read and reread the statistics on the racial disparity between the people who were evacuated and the ones who were left behind.

This was what the Maasai community member had been speaking of that day. And I had shot him down and told him it wasn't true. I'd told all these staff members that things were great for Blacks in America, that Black people didn't get hurt because of racism anymore. Now I understood his confusion over the seeming lack of value for Black life in a nation that asserts equality. I was ashamed of my country for letting me down and ashamed of myself for having painted an incomplete picture of America for my new African friends.

Compounded with this realization was another one: this major event had happened in my home nation, and I'd known nothing about it. Receiving this news weeks after it happened made me feel lonely and isolated. It struck me that I was so far away from everyone and everything I knew, completely disconnected. In many different ways, I'd never felt so detached from my country.

I brought the magazine to the Maasai village and sat down with several people, most of whom couldn't speak English. I paged through the magazine and explained, and some of the staff members translated. I said that there had been a big hurricane, and a lot of people died, even though they shouldn't have, a majority of them Black.

I admitted that I had been wrong, that I had led them to believe something that wasn't true. And I apologized.

The community member who'd yelled at me that day was extremely forgiving.

"We wanted to support you. We wanted you to explain this, so we could help you grieve because those are your people who are dead. Those are your people who are hurting. This is your community that needs help," he said. He then asked how they could pay a tribute to all the people who'd died.

I broke down in tears, moved that the Maasai cared about me and other Black Americans so much, even though we were so different and often misunderstood each other.

He said my tears were what had been missing that day, when I was confused and in denial. He told me that the village had cried for my people, and he offered asylum in his village for those of us who remained. And he declared that they would name the next baby born in the community Katrina.

ABOVE AND BEYOND the emotional and mental turmoil I experienced in Kenya, my physical safety was often jeopardized as well. The possibility of death was very real, and that awareness had been heightened by the thought of all the Black Americans who had perished in Hurricane Katrina. Those people had thought they were safe in their homes half a world away from me, yet they weren't. In contrast, the day-to-day reality of living in the African bush reminded me that sometimes safety is an illusion.

Everyone in my group was advised to double- and triple-check our tents for snakes, spiders, and other creatures that could inflict mortal harm. The same was true about our clothes—before we put on any item of clothing, including our footwear, we had to shake them out. Knowing that the closest medical facility was in Nairobi, six hours away—and already hyperaware of potential health issues because of my asthma and a prior bout of E. coli after our arrival—I was ritualistic about these safety checks.

One morning, we had to wake up extra early for a game count of zebras. Though I was a bit groggy, I shook out all my clothes and inspected all my gear. When I banged out my hiking boot, a palm-size scorpion dropped to the ground. It was in attack mode, its tail raised and stinger poised. It backed away from me, and I froze. All I could think was how grateful I was that in spite of being half-asleep and in a hurry, I'd taken time to shake out my boot.

As if daytime threats weren't enough, the nighttime presented what I perceived to be even greater dangers.

One night, I lay awake in my tent, every muscle in my body taught with tension. I really needed to pee, but I was terrified to venture into the darkness. I attuned my ears and could hear familiar sounds, like insects and owls. Yet some of the sounds were unfamiliar, and fear of that unknown made me hesitate, despite the fact that my bladder felt like it was about to burst.

I turned on my flashlight, and that's when I saw it: the silhouette of a big cat prowling around the tent, paws softly treading the ground. Thanks to our game drives, the shape of the sleek body had become familiar to me; it was a lioness. At one point, the lioness pressed her nose to the thin fabric that separated us. I figured that she was trying to discern whether her keen sense of smell was guiding her toward something worthwhile inside.

While I'd hoped to encounter lions, getting eaten by one in the middle of the night wasn't what I'd had in mind.

I was afraid to breathe, lest the shaky inhale and exhale be detected. Though I'd only read about lions in textbooks, I was convinced it was going to attack. Though I'm not a religious person, I started to pray, begging God to let me live.

The lioness circled the tent a few more times and once again pushed her nose against the fabric. Then, suddenly, she was gone. I didn't know if she had left the camp or was merely lying in wait to devour some poor college student who ventured to pee in the middle of the night.

Silent tears began to fall as I once again felt I'd narrowly escaped death.

I kept my flashlight on the entire night, and I didn't sleep a wink, my overfull bladder an uncomfortable companion.

By the time the sun rose and I could hear my teammates moving around the camp, I was filled with relief that the danger had passed. And that I finally could go pee.

After I had somewhat recovered from the experience, I got curious. Why didn't the lioness attack me? There was a thin piece of material separating me from her. It would've been easy enough for her to claw her way into the tent. I'd soon learn that, unlike bears—which will do anything to get at food, including tearing open a tent and attacking a human—most big cats don't waste their energy attacking things they can't see. While they might probe at something, like a tent, if they're not 100 percent sure that something is prey, they'll pass it by.

EARLY IN THE SEMESTER, I was recovering from a devastating *E. coli* infection that had completely wiped me out for a couple of days, and the high fever had made me delirious. Thankfully, I'd been prescribed some antibiotics and was soon on the mend—just in time for the class to complete the first activity on our syllabus: a hike into the Chyulu Hills, which would grant us an unparalleled view of the Great Rift Valley. This geographic feature inspired the scene in *The Lion King* where Rafiki the monkey holds up Simba, the future king, to be honored by the other animals.

Even though I'd never been hiking before, I wasn't nervous. I figured that I had all the gear—brand-new hiking boots, backpack, binoculars, water bottles—so I was prepared. Plus, it wasn't like we'd be climbing any mountains. At most, the incline we arrived at after an hour-and-a-half drive could be described as a hill. It had a trail that spiraled around and around, all the way to the top, which we could see from where we stood at the base.

Oh, this won't be too bad, I thought. We began our ascent in a single-file line, and I snagged a spot in the middle of the pack.

About ten minutes into the hike, I'd already dropped back to the end of the line because I had to keep taking breaks. I would've thought my asthma would do me in, or maybe even blisters from my hiking boots that hadn't been broken in yet. But it was my burning thigh muscles that forced my slower pace. It was a painful reminder that I never exercised. My asthma had gotten me excused from gym class during my public school years, and as a typical college student, physical fitness wasn't a priority for me.

Imposter syndrome made me want to give up. I wanted someone else to see my struggle and tell me I didn't have to keep going. I almost wanted my asthma to kick in so I had a legitimate excuse to stop.

But something in me knew that if I gave up now, I would keep giving up. And my dreams of becoming an ecologist and a nature host would never come to fruition.

It took me thirty minutes longer than everyone else to reach the top of the hill, and I hated every step. I had to continually give myself pep talks along the way. By the time I reached the top, I felt a mixture of triumph and humiliation.

From the top of the Chyulu Hills, we gazed across the vast Great Rift Valley, a place biologists note as the location of early human evolution. Our professors didn't rush us, and we stayed for over an hour. I understood why they'd brought us there so early in the program. The location not only gave us an incredible view of East African landscapes but also rooted us in time and place and purpose.

If my early human ancestors could evolve in Kenya, then so could I.

On the descent, I gave myself the credit that I'd denied myself just hours beforehand. This was a big deal for me. I'd pushed myself and did something physically rigorous for the first time. Something that people in my family didn't do.

For my entire life, I'd felt that I didn't have a place in nature, in the wilderness, that it was unavailable to me. I didn't see people who looked like me—whether a Black person, a woman, or a Black woman—represented as being stewards of the environment, being confident in and down and dirty with nature. The hiking experience was new and satisfying and made me feel proud of what I'd accomplished, and excited for what I could do in the future. Every step I'd taken up the hill that day was me taking one step closer to becoming the person I am today.

THE DAY AFTER THAT FIRST HIKE, our professors took us on our first game drive. Our camp was in close proximity to one of Kenya's most notable parks, Amboseli National Park, which was well-known for its large elephant population. Although we were headed to a national park, we weren't going as tourists. We piled into four Land Cruisers, armed with our cameras, as well as clipboards, notebooks, and an assignment to practice estimating distance and number of animals. Our professors said that by the end of the program, we should be able to eye a group of herbivores and jot down how many were present and at what distance from us—for example, thirty wildebeests one hundred meters away. Developing these skills would help us with the data collection we'd eventually do for our independent research projects midway through the semester.

We drove away from camp and down the dirt road that snaked through the Maasai village, called Kimana, that neighbored our camp. The drive was about forty minutes, and I peered out of the vehicle's dusty windows, watching the landscape turn from campsite to village to bush. Soon, not a shred of evidence of human life was within sight, and acacia trees dotted the landscape.

"The majority of wild animals live outside of national parks," one of our professors, Dr. Moses Okello, reminded us from the front seat. "You'll begin to see animals before we enter the park."

His timing was impeccable. As we rounded a turn, my classmate sitting next to me gasped.

"Elephants!" she yelled.

I stood up to get a better view, and that's when I saw them: a herd of elephants, about eight or ten of them, eating from an acacia tree. I couldn't believe how enormous they were.

"We will stop for more animals once we're within the park," said Dr. Okello. "If we stop for everything now, we'll never reach Amboseli!" He beamed with satisfaction, and I doubt he ever tired of this moment: the awe and wonder-filled eyes of green American students having their first sighting of iconic African wildlife.

During that afternoon at Amboseli National Park, I finally saw a lion. First, a lone male with a dark-brown mane. It was lying down, camouflaged in the tall tan grasses. Our driver, Maraka, an expert animal spotter, saw him first. He stopped the truck on the dirt road and instructed us to grab our binoculars. I pulled mine out and focused them where he was pointing. If it weren't for the lion's tail flicking every few seconds, I wouldn't have spotted him. When he came into focus, I gasped.

The lion must have been a good-luck charm, because the next hour was filled with numerous animal spottings: giraffes, zebras, wildebeests, gazelles, hippos, fish eagles, cheetahs, and more elephants. Some were active and excitable, like the antelopes that bounded away from us as our truck approached. Others were unbothered and conserving their energy, like the pride of female lions that barely glanced at us while they dozed in the afternoon sun.

My intention when I applied for the program was becoming a reality—I was living the life of a wildlife television host, sans cameras. But it was more than that. I was learning the ecology, the fullness of the landscape, and the details of the conservation issues, in real time and from the most authentic people. Our professors had been born in this place, then gone off to Europe and the United States to receive degrees. After graduation, they'd returned to advance wildlife conservation in their home country, while training the next generation of conservationists.

Being a nature-show host would always be one of my goals, but becoming a conservation scientist and ecologist—someone who worked to find ways to keep these wild animals, landscapes, and ecosystems thriving—was of even greater importance.

THE HIKE UP THE HILL to view the Great Rift Valley was neither the first nor the last experience in Kenya that would push me beyond the limits of what I believed myself capable of doing. Another one involved herding livestock, a new experience for all of the students.

Our team decided to pool our funds and buy a couple of goats and sheep because we were curious about what it was like to raise them. Every day we took turns herding them around the camp, and it proved to be extremely difficult.

One of the goats had been purchased specifically for slaughter. Toward the end of the semester, we'd kill the goat and have a ceremonial dinner, something that was an important part of the Maasai culture. When the time came, I was chosen as the student who would participate in slaughtering the goat.

I'd gotten the impression that the program staff wanted me to come away from this semester feeling a bit more African. That had shown up in many different ways, such as gifts of beaded jewelry or extra help with my Swahili. In this instance, that desire showed up as an invitation to slaughter a goat.

Prior to my trip to Kenya, I'd been a vegetarian for many years. But before I left, I decided to abandon my way of eating, since I didn't know what type of food I'd have access to while abroad. Even though I was fine with eating meat, I didn't want to kill a living, breathing creature. It wasn't who I was. I wanted to help save animals, not kill them. Yet it was an honor to have been invited to do it, and I knew I would offend my hosts if I refused.

The staff members helped me straddle the goat, and they showed me how to hold its horn with my left hand. They shoved a machete into my right hand and instructed me to slit the goat's throat. Hands gripping the horn, I pulled its head back and exposed its neck. I honestly thought the goat knew what was about to happen because it started bleating, and it sounded almost like a scream.

I placed the machete on the goat's neck, and though the machete drew blood, it didn't go deep enough. I ended up sawing the neck with the machete, and blood was spurting everywhere. The whole episode probably lasted about sixty seconds, but it felt like an hour to me.

Finally, one of the staff members took the machete from me and finished the kill, putting the poor animal out of its misery.

I walked away from the scene drenched in blood, even on my face and in my hair.

I cried for a long time afterward.

When I reflect on this experience, I recognize that it was appropriate. No one said killing an animal would be pretty or easy. You see the meat packaged in the supermarket, and it's so neat and clean. The whole process is sanitized. But killing an animal to provide a meal is meaningful, purposeful.

After that day, I vowed to never kill another animal, and I haven't. Even so, I'm glad I had the experience. It reminded me that I could do hard things—things I never thought I could do—and it further connected me to the culture I was immersed in.

*   *   *

A LOT OF MY TIME in the Maasai community was spent having reciprocal conversations, where I told them about the experiences of Black Americans, and they in turn told me about their cultural practices. One major difference I learned a lot about was how their approach to marriage contrasts with the American way. For them, marriage is an important part of the coming-of-age process, and those living a traditional Maasai lifestyle marry young, especially women. A man can have more than one wife, which is even expected if the man has significant financial resources. Wealth is typically measured by the number of cattle in a man's herd.

At the outset, my Western perspective dominated my outlook, and I judged these traditions. But during the semester, I came to recognize that cultural differences are just that: differences.

I became more comfortable with the idea of polygamy. When I met new people, I'd ask them, "How many wives do you have? How many children do you have?" Those kinds of questions were well received. But it was unacceptable to ask someone, "How many cows do you have?" because that's a measure of wealth. I learned this faux pas the hard way.

"How many cattle does your father have?" I asked a young staff member named Rauta.

"I'm not going to answer that question, Rae. It's like me asking you, 'How much money does your father make?'" he gracefully explained.

Livestock were herded all over the landscape in this part of southern Kenya, which was significant because the region was home to many species of wild animals too. Much of the study-abroad program was focused on visiting national parks and doing game drives and game counts. We also studied wild animals' movements outside the national parks, in the bush where Maasai communities and wild animals overlapped.

It didn't matter much if cattle shared space with other herbivores like zebra and antelope, but large predators like lions, leopards, hyenas, and cheetahs were often a threat. This was my first introduction to human-wildlife conflict, an area that I'd continue to study for many more years. And it was an issue that allowed me to deepen my relationships with the Maasai community, outside of our program's staff, in order to help solve the problems.

During the study-abroad program, one of the communities around us had an Eunoto ceremony, a coming-of-age ritual for young men and women. (In a Western context, we might compare it to a bar mitzvah.) The Eunoto takes place around age thirteen, and the ceremony is an exciting and beautiful event, full of dancing and music and feasting. It's also a fairly private event for the community, so it was generous of our Maasai neighbors to invite a group of curious American students to observe and participate.

I watched with my fellow students as girls were anointed with jewelry, praised and blessed by elders, and celebrated. Many Maasai *morans* (the Maa word for "warrior," a special age set of young men) performed dances that I'd seen in ethnographic videos back home. The best way I could describe it at the time, as I wrote in letters to my parents, was a dance of jumping. They sang and chanted while rocking back and forth and periodically jumped so straight and high it looked like an Olympic feat.

At the ceremony, I noticed that I received some special attention. I was a Black person from far away and without a rooted African identity. I also was young and unmarried, which they considered unusual for someone my age—the same age as many of the Maasai leading the ceremony. Regardless, I wasn't comfortable striking up a conversation with any of the young Maasai men or women. I was feeling shy and perhaps a bit overwhelmed. And maybe I was slightly jealous that these people, many of whom were my age, had such a strong connection to a tradition that was thousands of years old. We were all Black, yet I felt like more of an outsider than my white classmates.

A couple of weeks after the Eunoto ceremony, we had some downtime in the camp. I was sitting outside my hut, gathering clothes and trying to muster up the motivation to hand-wash my laundry. I heard a motor scooter buzzing up to camp, an unusual sound.

A few minutes later, a student named Bridgette, whom I had become close friends with, trotted over to me.

"Rae, there's someone here to see you," she said.

My first feeling was a bit of panic. *That's impossible*, I thought. *The only reason anyone would come here to see me, all the way from the United States, is if something horrible has happened.*

I stood up in a daze, trying to steady myself to receive the devastating news.

When I walked to the entrance of our camp, a couple of the male staff members were speaking with a guy from the village, whom I didn't recognize. My initial distress was replaced by curiosity and confusion.

The stranger was attractive and young, probably my age or a little older. He had smooth dark-brown skin and a shaved head, which was common for his age group. Instead of a shirt, he wore vibrant cloths draped over his chest like a sash and around his waist. A machete was tucked into the cloth that was wrapped like a skirt. He had beautiful traditional Maasai beads around his waist and a few around his neck. His garment revealed toned biceps and a chest chiseled from a life of activity.

He sat on his motor scooter with the confidence of an influential person in the community. Sunglasses shielded his eyes from the glare of the midday sun as he flashed a bright, white-toothed smile while speaking with the staff members. Spotting my approach, one of the staff motioned for me to join them.

"Solati would like to meet you. He is from the farther village," he said.

I'd been having such a great time getting to know members of the Maasai community and having a sense of exclusive entry because of my Blackness. So I just figured I was receiving a special introduction to a friendship because I was the Black student, who also happened to be especially interested in Maasai culture.

"Hi, nice to meet you. My name is Rae."

"It's nice to meet you. My name is Solati," he said in perfect English.

An uncomfortable silence that seemed to last forever hung in the air.

Eventually, the two staff members exchanged some words with Solati in Maa. One of the staff members finally said, "Okay, Rae—you can go now."

I walked back into the hut that served as our classroom, where two of my classmates had been eavesdropping.

"Oh my God, what's going on? What was that about?" they asked.

"There's this guy from the next village over. Did you see him? He's cute. And he wanted to meet me."

"Maybe he wants to date you. Maybe he heard about you and heard you are beautiful and wanted to come see for himself," my friends said, hyping me up.

I was happy to entertain that fantasy. The idea of having a Kenyan boyfriend had never occurred to me, but it was fun to think that could happen. Despite my excitement over that possibility, months went by, and I never heard from or saw Solati again.

Before we knew it, the semester was nearly over. A day was scheduled for all of the staff members to drive us from the camp to the airport in Nairobi. Most students were flying out of Nairobi that day, overnight to Europe, and then from Europe to the United States.

Several weeks prior to that, I'd hatched a master plan with my friend Bridgette. We were hooked on Africa and didn't want to fly back home to the States right after the program ended. Even if it meant missing Christmas with our families, we wanted more time to backpack around Kenya. We'd written letters to our parents, who changed our plane tickets, giving us two weeks of freedom on the continent.

Because the program was ending, Bridgette and I needed to pack our things and leave camp along with the rest of the students. I had way more than I could carry for two weeks, so I hand-washed my clothes and left them neatly folded for the staff members to either keep for themselves or distribute to community members. I also left books, supplies, bags, and everything that wouldn't fit in my backpack. My burden was lighter than ever, and I was ready for adventure.

On the morning we were preparing to leave, once again the sound of a motor scooter filled the air. And there was Solati.

As before, he was wearing sunglasses and beautiful cloths draped over his body. He was also wearing a leather motorcycle jacket and sturdy sandals. I was busy hauling my belongings toward the trucks that would take us to Nairobi. I felt myself smile on the inside, anticipating that Solati had come to say a special goodbye, producing another flirtatious moment that I'd be able to replay in my head while I daydreamed during the long drive.

But this time, there was a palpable seriousness that arrived with Solati.

"Rae, you're going to be staying here," said one of the staff members who'd introduced me to Solati.

"Oh, I wish I could—I really wish I could. But I have to go back. I promise I'll return. I can't wait to come back to Kenya."

The staff member's brow furrowed.

"No, Rae, you're going to be staying here. You're going to go with Solati," he told me.

I stared blankly at him. Then I glanced to Solati, whose eyes I couldn't see behind his sunglasses, and back to the staff member.

"Solati will take you as his wife," the staff member said.

Tension coiled in the pit of my stomach.

"Excuse me, what?" I looked around for American friends who could help me sort through this situation, but no one was in the immediate vicinity. The staff member calmly laid a hand on my shoulder and stooped down to make eye contact with me. He spoke gently.

"This is a really big honor. He is the son of the chief of the community. He has never married before. You will be his first wife, which is the biggest honor."

I'd only been living in Kenya for four months, so I was far from an expert on Maasai cultural traditions. But I had been listening and learning, and I understood that according to tradition, this was the opportunity of a lifetime. This man was wealthy. He owned acres and acres of land, with many cattle. He was prominent. Essentially, the English translation for his title was Prince Solati.

My mind swirled, and my whole body felt hot. I acknowledged that this marriage offer was a high compliment to me. I was considered someone worthy of being with this prince. But this wasn't a marriage proposal—Solati had arrived at the camp to take me away. Somehow, the marriage had already been arranged.

I backed away while saying, "No, no, thank you. I have to go back home to the United States, and I will return to Kenya. But I'm not going to get married."

The staff member who seconds before had been so gentle became upset.

"You don't understand. You have to get married—he's chosen you. You're not going to Nairobi. You're going to go with him."

In that moment, I felt afraid, panicked that I might need to escape these men.

Today, I would've handled the situation in a completely different way. I would have rung the alarm and gotten the help I needed from whoever was at the camp. But back then, I was embarrassed and I felt ashamed. Perhaps I'd inadvertently brought this on. Perhaps I'd been too fun and congenial and flirtatious. I blamed myself for this misunderstanding, something that American women have been socialized to do.

"No, I'm not doing this. I'm not going—I'm going to Nairobi. And I'm going to travel with Bridgette, and then I'm flying home," I said. My chest felt tense, as if something were gripping it. I then turned and walked away from the men.

I had spent the past few months studying wild animals, including their fight-or-flight responses. The moment I felt caught in this trap, had I fought by putting my foot down and saying no? Or was I taking flight by walking away? I needed support, I needed protection, and I needed help. But once again, shame overtook me in this vulnerable moment.

I returned to the camp, where all the students were hustling to pack up. I didn't say a word to anyone. I went to finish packing, and while I was doing that, Rauta, whom I trusted, came to check on me and explained what had happened.

Some of the staff members had determined that, although I had a father in the United States, he couldn't be reached. Because the father is the one who gives permission and then arranges the exchange of a dowry, I needed someone to be the proxy for my dad. So the staff member who was there and speaking to me had taken on my father's role in the betrothal. He'd spent months negotiating the price of my marriage.

There are parameters around how a young Maasai man chooses to marry, deeply rooted in tradition. I hadn't realized that what had happened months earlier wasn't a cute flirtation: it had been the beginning of a formal marriage proposal that I wasn't privy to.

For months all of these men had been talking about me and deciding things for me, creating this idea of what would be best for me and for Solati. To this day the situation elicits a bizarre feeling, caused by the combination of viewing the betrothal both as a huge honor and as hugely offensive.

Hours passed, and no one bothered me. I gathered my stuff and threw it in the truck.

The entire duration of that six-hour drive to Nairobi, I pondered what had happened, still steeped in shame. I had wanted the end of my time at the camp and with the community to be celebratory, full of triumph and positivity. I wanted to believe that I was a different person, perhaps less American and more Kenyan than when I'd arrived. Yet that desire was being challenged by this major and uncomfortable misunderstanding.

We arrived in Nairobi, and I told myself to shake it off and enjoy my classmates' company. Most people didn't need to be at the airport for hours, and our drivers had found a restaurant where we could eat and relax until it was time to say goodbye. We ate *nyama choma* (roasted meat) and drank cold beers. The latter was a refreshing delight after months without refrigeration.

As we were sitting there, Solati rode up on his motor scooter.

Instead of the traditional Maasai clothing I'd previously seen him in, he wore jeans and a T-shirt under his leather jacket—like any other city guy zooming around on a motor scooter. He hopped off the scooter and placed his helmet on the seat. His eyes caught mine, and I heard him say my name for the first time.

"Rae," he called out, pointing to me and curling his finger, signaling for me to come over. "Rae."

Women and girls in the United States are taught to think of gestures like these as romantic: a man travels a great distance and makes a grand gesture in order to win your hand in marriage. Something like that is out of a fairy tale. Yet in reality, the scenario was frightening. I hadn't told any of my classmates what was going on, and once again, I opted to navigate the situation on my own.

Solati beckoned me to come over to speak to him. After I walked over to him, we had an almost identical conversation to the previous one. Except this time, he spoke directly to me. I could tell he was confused. He didn't seem angry, which was a relief, but he did seem somewhat desperate. I knew he wasn't going to hurt me. My guard came down a bit, although I didn't have the words to say anything different from what I'd already said: "No, I'm not going to get married. I can't get married."

"Why not?" he asked with pain in his eyes. He said that he'd done all the things required of him, all the things that he knew to do.

I could feel the tragic nature of this moment. It was possible that he'd begun to love me—or at least the idea of me—and begun envisioning our life together. Wedded bliss with me, his first and most honored wife. And now that dream was swiftly unraveling.

I racked my brain, searching for a reason that might make sense to him. What I landed on was something that wasn't entirely my truth, but I felt he could understand it.

"I do have a father, and he has not been contacted. So I need to ask my father if this is okay," I explained.

I wish I'd been strong enough to say, "I don't want to do this. I don't know you, and this isn't how it's going to work for me." But I wasn't, and therefore I resorted to an excuse.

Solati was resigned, and his shoulders hunched. I could tell that he felt like he had lost so much.

I gave him my mailing address, so he could send me something if he wanted to. And I gave him my email address, in case he ever found himself at an internet café. I regret that I left him in limbo, wondering if the answer would eventually be yes. That somehow, I would come back with my real father's blessing.

But I never heard from him again. I don't know if that's because he figured out that I was never coming back or if he wasn't able to contact me.

I said goodbye to him, and he hopped on his motor scooter and sped away. Although I didn't want to go with Solati, I felt a possible future was just a ride away with him—a life as a Kenyan, fully committed to this place and this culture.

* * *

I OFTEN HEARD the cliché that when a Black American visited the African continent, being surrounded by similarly dark-skinned people made them more confident and empowered in their Blackness. This confidence then led to the stripping away of any white beauty standards that American society had forced onto them—whether that be hairstyle, makeup, or fashion.

When you're socialized in American culture, especially as a woman, you learn that being beautiful and being Black are often considered mutually exclusive. Throughout my early life, I struggled with this tension in various ways. Of course, I couldn't do anything about the color of my skin, but I could do something about my hair. Deep down, I didn't necessarily want to change my hair texture, yet adolescent and then teenage me desperately wanted at least one of my physical features to be mainstream and acceptable, and thus considered beautiful. So from 1998 to 2005, I used a relaxer.

Unlike me, my mom had always been strong enough to withstand those kinds of societal pressures. When I was growing up, she'd often tell me how my grandparents used to force her to chemically straighten her kinky hair when she was a teenager. When she was eighteen years old, in a huge act of defiance, she chopped off her "gorgeous" straight hair, leaving less than an inch of nappy black coils.

The way she tells it, she was damn near kicked out of the house. My grandparents knew this choice was the beginning of the end of my mom allowing them to manipulate her body and choices. Although they were well-meaning in many ways, my grandparents didn't trust that my mom could successfully navigate through white America as her authentic self. They were wrong. And since that day, my mother and her natural hair texture have forged a unique, beautiful path through this world.

After the drama surrounding relaxers in my mom's life, you can imagine how horrified she was when I, at age thirteen, asked if I could get one. I know my mother struggled with parenting in this moment, because she had previously put her foot down when it came to certain beauty rituals that she'd deemed inappropriate, whether because of my age or the reality that they stemmed from unfair beauty standards that emerged from a toxic part of white America. Yet perhaps she realized that, much like her eighteen-year-old self who'd cut off all her hair, I was going to find a way to do what I wanted with my physical appearance, impossible white beauty standards or not. While I wouldn't say she supported me, she did allow it. Even at that young age, I knew she was afraid of creating a rift between us, similar to the resentment between her and my grandmother. And I was willing to disappoint my mom because acceptance from my peers was a much higher priority for me as a teenager.

Upon seeing me with my relaxed hair, my grandparents' praise was effusive. Suddenly, I was much more beautiful. I was a "woman." I was the refined, sophisticated grandchild of their dreams.

My new hairstyle made a big difference at school too. I don't think any of the white kids noticed or cared, but the way the cool Black girls finally saw me as worthy of being spoken to, and the way the Black boys finally looked at me like an eligible young woman, gave me a sense of belonging that I'd never experienced. So I continued to use the relaxer throughout high school and into my college years.

Before I left for Kenya, I had to decide what to do with my hair. Keeping up a relaxer was out of the question, so a few days before we departed, I went to the salon and got my hair styled into hundreds of gorgeous little box braids that cascaded down to the middle of my back. I loved them and wondered why I'd never considered this style before. It was flattering, protective, and fun, and it helped me set out on this new adventure feeling like my appearance was at its best.

After Bridgette and I parted ways with the group, we walked around Nairobi. We had no plan and figured that we'd either camp or stay in a hotel.

For the first time in my life, I was completely untethered. To celebrate and signify this rite of passage, I found a hair salon and cut off my braids at the roots. With each snip of the scissors, I felt more and more free. Free of expectations, whether societal or self-imposed. Free of notions of what I "should" look like, as I blended in so beautifully with the Kenyans around us.

Bridgette and I got connected with an orphanage for Maasai children, and we ended up volunteering there during our extended stay. While we might have thought we'd spend our days changing diapers, warming up bottles, and reading to toddlers snuggling in our laps, that wasn't what we experienced at all. Our days were spent doing hard manual labor, like chopping wood, cleaning, and cooking. Even though the work was strenuous, we took it all in stride, and when we left, we felt a tremendous sense of accomplishment. And I was grateful to have had an opportunity to serve these Maasai children, born in a community that had given me so much.

I arrived home on Christmas Day in 2005. The best gift was seeing both my parents when they picked me up from the airport. Asa was also home from college, and the four of us had a joyful reunion.

It felt wonderful to be home, but in my very being, I knew I'd one day return to Kenya—and when I did, it would be on my own terms.

*   *   *

ONE MIGHT THINK that a semester spent in the wilds of Africa would have solidified my belief in myself and my path toward becoming an ecologist. Yes, my passion had been sparked, but returning to the environmental science classroom sank my self-esteem.

At the end of a session of an environmental policy class, which was taught by a grad student named Paul Hirsch, he asked me to stop by during his office hours.

"You know, part of your grade is based on class participation, and you never raise your hand or say anything. Are you all right?" he said to me in his office later that day.

"I feel like everybody in class already knows everything, and I'm learning this US environmental policy stuff for the first time. I'm self-conscious about participating and being wrong. I don't know anything," I told him.

"I'm going to go out on a limb here, and I hope that's okay. I've noticed that you're the only Black student in our class and in the department. My girlfriend is doing her PhD in economics. She's the only Black student in her program, and it's really hard for her. Is that some of what's going on?"

I burst into tears right there in his office—not out of shame but because I felt seen. Someone with power and privilege had acknowledged the problem without me having to articulate it.

"I'm sure you know lots of things," he reassured me. "And you deserve to earn your participation grade. Let's try this—tell me what kind of environmental policy you *do* know about."

"Well, I spent five months living in Kenya studying wildlife management, so I know all about what's going on with wildlife conservation in East Africa."

"See! I bet none of those boys in class have experience with that!"

He and I then created a plan for me to be the guest lecturer for two class sessions. I'd create PowerPoint lectures about East African environmental policy, and I'd give reading assignments and homework.

In a brief yet sacred moment, my professor reset my course: he empowered me, and he helped me see how to empower myself. I went home, crafted my two sessions of guest lectures, and soon was standing in front of a class of my mostly white male peers, teaching them what I knew best from my transformative Kenya experience. This first taste of professorship reinforced my desire to share my expertise with others, to inspire and challenge them.

This experience also allowed my "double life" to become singular, as the strong, wild Rae of the outdoors and the meek, shy Black girl of academia merged.

I finished my environmental science degree with high grades and even graduated from Emory a semester early in December 2006, ahead of most of my friends and all the other environmental science students. It was a powerful end to my collegiate experience, and I felt more ready than ever to immerse myself in wildlife conservation work, this time in the professional arena.

I applied for various jobs and was elated to receive an offer from the World Wildlife Foundation at their headquarters in Washington, DC. It was an administrative assistant position, but it would get my foot in the door and situate me in a city where many environmental organizations were headquartered— including one that I had always dreamed of working at, the National Geographic Society.

*   *   *

TWO YEARS LATER, in 2008, I was still working as an administrative assistant at WWF. Although the work itself didn't interest me, the surroundings inspired me. The conservation scientists who worked for the organization were people I wanted to emulate, and the organization's general mission aligned with my own goals of working collectively to save endangered species from extinction.

I spent most of my days making coffee for guests, taking notes in meetings, and filing receipts from my colleagues' travel. Scientists dropped receipts onto my desk from their latest excursions, such as studying tigers in South Asia. I'd take notes in meetings where conservationists from WWF's international offices would call in to talk about the state of mountain gorillas in the Democratic Republic of Congo. As I observed from the outside, I constantly asked myself what all these people had that I lacked. It wasn't passion—it was graduate school. I already had a bachelor's degree in environmental science, but to reach my goal, I'd need to further my education.

For several months I studied for the Graduate Record Examinations (GRE), and I applied to all my colleagues' alma maters. The programs ranged in location from nearby state schools to high-tier departments in California's UC system, as well as several Ivy League programs. The last time I'd applied to college, I was a high schooler grappling with Cs and Ds on my transcript, with a fire inside of me to become independent and find my passion. This time around, I had emerged from Emory with an impressive grade point average and strong recommendations from my professors. I also had a clear picture of what kind of professional I wanted to be and how serving the planet would answer my purpose. Yet I believed it was too lofty a goal to be admitted to Columbia or Yale. Asa was the super-smart, Ivy League sibling in the family, not me.

But the day prior, I'd been stunned to receive an email from Joanne DeBernardo in the admissions department at the Yale School of Forestry and Environmental Studies (today named the Yale School of the Environment). She introduced herself and informed me that there'd been some kind of hiccup in their system, so they'd be calling applicants. I could expect to hear from them the next day. The personal touch felt promising, but the email didn't indicate one way or another what the decision would be. And the self-doubt in the back of my mind reminded me not to get my hopes up.

After I read the email at work, I texted my mom to tell her that decisions would be announced the next day and then headed over to tell my coworker Nora. Nora, a white woman from Cleveland, Ohio, was one of my best friends from Emory, and she also happened to land in Washington, DC, after graduation. After struggling to find an entry-level job in her preferred field of international development, she'd found an administrative assistant opening at WWF. Had the hiring managers known we were the type of best friends to blow off work to sit together and giggle for hours each day, she likely wouldn't have gotten the job. But during the hiring process, we were subtle about our relationship and professional, and she was hired. Her first week was the most fun I'd ever had at the office, and months later, it was obvious to me that having an in-office best friend was the ideal way to navigate these early professional years.

The day I received the email from Yale, I told Nora the new time line, and she squeezed me tight in a hug. Her optimism far exceeded mine.

Part of me still didn't believe I could be admitted into a school like Yale, but they had taken the time to reach out to me, which teased the eternal optimist in me. Though I often stress about my present circumstances, the future always looks bright to me.

The next day, I knew I wouldn't be able to focus on work. I'd brought my lunch so I wouldn't have to leave my desk to eat, and I asked Nora to wait in my chair the couple of times I'd raced to the bathroom.

Finally, the call came.

"Hi, is this Rae? It's Joanne DeBernardo from Yale."

By some miracle, I managed to answer her affirmatively without my voice quivering.

"Rae, congratulations—we're happy to offer you a spot in the class of 2010!"

My hand flew to my mouth to keep myself from screaming. "Thank y—"

"And I have more good news," Joanne interrupted. "We are also happy to offer you a full scholarship to attend our program, and we hope this makes a big difference in your decision."

"Are you serious?" My loud voice carried across the office. So much for being subtle. "Oh my gosh, thank you so much!"

My head spun. Was I being punked? Yale was admitting me for a master's program? And with a full scholarship?

"Rae, let all of this soak in, and you can expect an official offer in the mail within a week. Again, congratulations, and give me a call if you have any questions in the meantime."

After we hung up, I looked around the office. No one seemed fazed by my outburst. The researcher whose desk was closest to mine had his headphones plugged into his computer and was working away, oblivious to the fact that my entire future had just changed in an instant.

I pulled myself together and strolled over to Nora's desk, which was empty. Without hesitation, I grabbed my purse and ran downstairs, out the front doors of the WWF building, phone in hand. I dialed my mom's number.

"Mommy?" I said, pressing the phone closer to my ear, willing the technology to cooperate.

"Toot, can you hear me?" she asked.

"Mommy? I think our connection is bad." After several frustrating delays because of a bad connection, I was finally able to share the news: "The people from Yale just called. I got in!"

My mom screamed into the phone. Though she always shares in my joy when I'm excited about an opportunity, she usually doesn't get so animated. It felt wonderful to share that moment with her, someone whose love for me wasn't tied to my performance or accomplishments, who had always believed in me no matter what.

WITH THE NEWS of my acceptance into the Yale School of the Environment, I was happy to file my last expense report, sit in my last team meeting, and say so long to my colleagues. Part of me hoped that earning a master's degree would bring me right back to WWF, where I could work as a star conservation scientist.

Although I generally had a wonderful and inspiring tenure at WWF, the lack of Black folks in leadership discouraged me. Black women dominated the administrative positions, and almost every janitor was a Black man. I used to deliberately go downstairs to the maintenance department instead of calling them on the phone because it gave me an excuse to hang out with Black folks at my workplace while still being on the clock.

When it came to the scientists at WWF, and those who were designing and doing conservation work, almost everybody was white. I had friends in many of DC's other environmental non-profit organizations, so I knew this issue wasn't unique to WWF. It was a pattern that many people felt was wrong, but no one seemed willing to make meaningful change to address it.

I had no intention of waiting to be given an opportunity to grow into a science position at WWF. I took it upon myself to become undeniably qualified, and I had the energy and motivation to do it. I daydreamed about the optics of the young Black administrative assistant returning years later, but in a leadership position—and with an Ivy League education to boot.

Chapter 4

Off I went to New Haven in 2008. I knew exactly what I wanted to get out of my master's program: to enter as a student and emerge as a full-fledged scientist. I envisioned my wildlife conservation experience would entail returning to East Africa, spending more time in the field, and helping endangered species avoid extinction.

Thankfully, the advisor I was assigned was a perfect fit. Dr. Susan Clark, a large-carnivore expert who had spent decades studying human-wolf conflict in the American West, believed in me and my passions. Right away, she tasked me with finding a project in East Africa that would help move me closer to my goals.

More important, Dr. Clark was able to deeply empathize with the reality that my identity influenced my experiences in science. She was a transgender woman who had transitioned a few years earlier, so her professional community was still learning of and accepting the change. Being her unapologetic, authentic self in the wildlife conservation space took bravery, and she chose to speak openly and often about the trans experience as part of her advocacy work. From our first moments working together as advisor and advisee, we bonded over the lack of diversity in our field and our dedication to making positive change for those who would come after us.

One day, I nervously told Dr. Clark that my dream project would be to study lions in Africa.

"If you want to study lions, I have the perfect person for you to work with in Tanzania," Dr. Clark said. "I'll email Laly and see if she can take on a graduate student."

After that meeting, I went online to learn everything I could about Dr. Laly Lichtenfeld. She was an alum from the same Yale program I was currently in and was a National Geographic Explorer. She'd dedicated her life to big-cat conservation, specifically lions in Tanzania. She was even in the process of building a camp for her organization outside Tarangire National Park in northern Tanzania.

Dr. Lichtenfeld and I started an email correspondence, and within days we hatched a plan for me to spend the second half of my summer working with her team on a spatial analysis, where we'd study patterns of human-lion conflict in the region. That same week, I connected with a former colleague from Washington, DC—a man named Kaddu Sebunya, who held a leadership role at the African Wildlife Foundation. He connected me with a group of AWF researchers in Tanzania that was also working on a lion project and could use the help of a graduate student. I would spend the first half of my summer with them and the second half with Dr. Lichtenfeld's group.

Between the support of my advisor, the excitement of my upcoming summer research, and the widespread energy of the students I shared space with every day, the semesters flew by, and I excelled in my schoolwork in a way that amazed me. By the time classes wrapped up for the year in May 2009, I was no longer the Rae in Kenya who'd feared being devoured by a lion. I had grown and matured into the Rae who was ready to spend an entire summer studying lions.

* * *

I ARRIVED IN ARUSHA, TANZANIA, on a Tuesday afternoon, after a six-hour bus ride through Kenya. We drove from the Nairobi airport through the towns of Kitengela, Kajiado, and Namanga, and then past rural Maasai villages. Seeing the familiar dwelling places felt like a homecoming.

In the bush, June is the beginning of the dry season. Due to climate change, Kenya had been suffering unimaginable droughts. Roads were barren and dusty, and everything looked brown. Interspersed throughout the sparse landscape were acacia trees and wildlife, zebras and gazelles hiding behind bushes and branches.

As we neared the Kenya-Tanzania border, I grew nervous about the approaching checkpoint. For my East African colleagues, crossing was usually effortless. However, as an American, I would probably be hassled. Americans in that part of East Africa often traveled with a sense of entitlement, which inclined the border-patrol agents to give all Americans passing through a hard time. Additionally, traveling alone in that part of Africa as a Black woman was fairly taboo at the time, so people were often confused about who I was and why I was there. Not adhering to social norms definitely posed challenges for me.

Sure enough, the immigration officers insisted I pay a hundred-dollar fee and proceeded to grill me about being an American, having a hyphenated last name, possessing a Kenyan student visa from 2005, and being Black. After they finished their equally uncomfortable and unnecessary interrogation, I was finally allowed to cross the border into Tanzania.

I spent my first several days in Arusha defeating jet lag, meeting some of my new colleagues, registering for research permits, and gathering supplies with the team in advance of our fieldwork.

Tanzanians speak a subtle subdialect of Swahili. Despite being unfamiliar with the intricacies of the regional speech, I was happy to know as much universal Swahili as I did. It would be a huge advantage, because I'd be spending time in places where few people spoke English.

The pressure to be fluent was ever present, as the Tanzanian men were particularly demanding. They flooded my days with questions, insisting that I explain who I was and why I wasn't behaving like the women of the area. I was both targeted and shunned for doing things I did every day in America: Walking by myself. Showing my face in public during the day, when women were typically at home cooking and cleaning. Eating in public. Interacting with strangers. At times, I found camaraderie with people I met, such as those who wanted to ask me about America—especially about Obama, who'd just won his historic presidential election—and why I was in Arusha. Unlike the complicated misunderstandings I'd had during my first trip to East Africa, this time I knew how to be properly assertive and avoid most issues.

After having a bit of cultural immersion in the city, my research team and I began our journey into the bush. We traveled for a full day to reach the camp. It was a rugged, scrappy camp in the bush, outside of Tarangire National Park, about a six-hour drive from Arusha. Our research facilities consisted of three huts, one of which had a satellite dish attached to the roof, which provided the station with a slow internet connection that could be used in an emergency. It was a far cry from the landscaped brick and ivy of New Haven.

Each morning, over a cup of chai—the sweet, milky tea that commonly serves as breakfast in Tanzania—I'd observe zebras feeding on grass, with an occasional interruption from hyenas, leopards, lions, cheetahs, and elephants. Despite the majesty of this wildlife, there were plenty of threatening animals too. Scorpions, tarantulas, vampire bats, and venomous snakes lurked in the dark corners of our huts. When the sun set each night, I prayed to wake up the following morning free of a mystery roommate or an indistinguishable bite.

A young, less experienced version of me might have been intimidated by how much we were roughing it, or by how few emergency services were available. However, in this moment in my life, this type of experience was exactly what I wanted. I was part of an exclusive team doing real scientific work. It was a wonderful place to call home for that summer.

The team of four consisted of the most impressive people I'd ever met. Ifura, our fearless leader, was a woman in her late twenties. Steven was a general assistant who helped in all kinds of ways. Eddie was a fix-it man. And Peter was second in command to Ifura. He was another up-and-coming scientist, a super-tall guy. He came from a nearby Maasai village, so he also helped us relate to the community.

Despite the bush's intoxicating beauty, its charm was not without trials. Water, due to the drought, was often scarce. As difficult as this was for us, it greatly impacted the region's wildlife as well. Nightly, we found ourselves chasing various animals out of the camp and away from our water supply.

One night, as we conducted a routine sweep, we noticed a baby elephant that had wandered astray while looking for water. At 2:00 a.m., our research team banded together to chase the baby away before all our precious water was guzzled down. Of course, babies are seldom far from their mothers, and this mama wasn't too happy with us. The following morning, we woke up to a crushed Land Cruiser and an empty canteen.

Fending off marauding elephants made for good stories to write home about when I was able to get to the city and access the internet, but this wasn't my main purpose for being in Tanzania. I was there as a research assistant with my new AWF colleagues, to study how certain landscape features influenced lion predation of Maasai livestock. We'd be tracking lions and seeking patterns in their movement.

Daily, most of our team would pile into our junky, clunky field vehicle and head out into the bush to look for lions, which we tracked with radio telemetry. The previous summer, the research team had outfitted one of the lionesses in the pride with a radio collar around her neck. A team member would stick the antenna out the truck window and hold it at different angles to try to pick up a radio signal. The sound began as a faint beep that would grow louder as we got closer to the collared lioness.

To identify the optimal habitats for lion conservation, we had to determine both where they went and where they avoided. Our work also helped us communicate with the resident Maasai people, who are pastoralists, so they could make cattle-herding decisions based on where they knew the lions did or didn't roam.

On a typical day, we'd spend hours tracking down the lion pride and then observe them for about an hour. Ideally, they'd do something, so we could document patterns in their behavior. But I'd learned many times that lions rest for about twenty hours a day. So the data collection often looked like us resting in our car while the lions rested with their pride—humans and wildlife reclaiming their time and energy in a world that wanted us to be in constant motion.

I'll never forget the time we found a huge adult male lion with a glorious, full mane—like Mufasa from *The Lion King*. He had just killed a zebra, probably around dawn. He was lounging next to the kill, his belly engorged, almost like when a cartoon character overeats to the point that they can't move. The lion was panting because he was trying to digest, and all the fur on his face was saturated with blood. It was amazing.

While we watched him, some other lions came by, which was exciting because this was a solo male that wasn't attached to a pride. He was so full that, although he growled a bit, he couldn't fight the other lions to protect his kill. *I definitely didn't see* this *on nature shows,* I thought to myself. The idea suddenly dawned on me that perhaps not everything I saw on television—even on my beloved nature shows—was always the full story.

LEARNING TO TRACK LIONS only took me a few days, but identifying the individual lions took longer to master. Ifura taught me that the key is to study the lion's whisker pattern: the spot on the lion's face where the whisker sprouts from its cheek has a tiny freckle, and every lion has a unique pattern of those freckles. "It's like a human fingerprint," she said as she instantly became my favorite scientist I'd ever met. "You can identify an individual lion based on its whisker pattern. It's super cool—try it!" She reached into the truck's glove compartment and handed me blown-up, laminated pictures of the individual lions that were part of the study.

We'd peer through our binoculars and try to count the whiskers on a lion's cheeks. If we could identify the lion, we'd then document data in a notebook. Over the summer, we mostly focused on three lion prides, a total of twenty-something animals. The majority of them had already been given names, many of which were Christian, like Mary and Joseph. However, most of the lions in the pride I studied had lovely Swahili or Maasai names. One of them was Nashipae, which means "happiness" in the Maasai language.

When new lions entered the pride or passed through the region, we got to name them. Often, we stuck to numbers and codes, such as "Green79," for the color of the radio collar we attached to the lion and the last two digits of the code we'd assigned it. Since the researchers got to name the lions, their names varied from creative to boring, depending on the researcher's personality.

"Can I name the next one?" I asked Ifura one day.

"Umm . . . maybe," she replied.

Had I sounded entitled or crossed some kind of line? Maybe I hadn't been there long enough to earn the privilege. Still, I hoped that before I left Tanzania that summer, I'd be able to name a lion.

One morning as we were preparing for our day of lion tracking, a young Maasai boy raced into our camp. His body heaved from his deep, rapid breathing.

"Michael Jackson is dead," he announced.

Impossible, I thought. Michael Jackson wasn't the kind of person who up and died. He hadn't even been sick. Was it an accident or something?

The boy explained that someone had visited the bigger village a few hours away and heard an English radio channel say that Michael Jackson was dead. The Maasai perceived this as crucial and sensitive information because they knew my middle name is Jackson. Multiple times I'd explained my name to them, including the fact that, unlike everyone else in the community, my name had no significance—it was just a name. Their concern was that the King of Pop must have been a close relative of mine—perhaps someone I was named after—and I'd therefore need to go home to be with my family.

"No, no, no. No!" I said in English, Swahili, and Maa. "He is not dead. It must be a rumor." Similar to when I'd been told about Hurricane Katrina, my mind and heart refused to believe this devastating news from home. I began trying to explain the concept of tabloid magazines to the boy. I told him how they printed false claims about celebrity deaths all the time. I realized how ridiculous that attempt was, so I stopped myself and rushed him back to the village. I thanked everyone for their concern but said it was a misunderstanding. And, once again, I insisted I was not related to Michael Jackson.

When I arrived back at camp later that afternoon, I was met with concerned looks from my East African colleagues. They'd had access to the radios all day, and the news was true after all. The tears that welled in my eyes and streamed down my face were partially due to grief for the tragic, unexpected death of one of my favorite artists—a larger-than-life man who'd seemed immortal. But the tears also belied a feeling that was beginning to surface, one that caught me off guard: I was homesick. In a land of beautiful Black faces and traces of my ancestry, I longed for the Black American community.

Perhaps one of the most surprising results of exploration is that it makes you appreciate home.

The next week, we discovered a new lion in the region. A juvenile male was dispersing through the plains, likely hoping to establish new territory and join a pride. I was given the privilege of naming this one, and I called him Michael Jackson.

WHEN YOU WORK for months in the field, it's natural to form bonds with the animals you're studying. You don't become a wildlife ecologist unless you already have an innate empathy for animals, not to mention the animals in your care. Naming the lions further cements that bond. Studying the prides that summer, I grew to appreciate their magnificence, their grandeur.

Each one played an integral role in their pride, completely different from the next. Beautiful Nashipae, for example, chased prey to the center of the hunting group for larger lionesses to capture, and she nursed and protected her cubs.

As our research continued, we heard news of lion attacks in a nearby Maasai village. Three Maasai boys herding livestock were attacked, killed, and eaten by two lions. In retaliation, the Maasai began killing the animals, outraged by the unjustified attack. As fellow human beings, we were heartbroken about the boys' deaths. And as researchers, we were unsettled by the Maasai's actions.

Animals, like humans, are complicated. They understand community, and they understand kinship. I was devastated to learn of their attacks on the hospitable Maasai with whom we were staying. How could these animals I have loved cause such harm? And how could the Maasai, with whom I had built strong bonds of friendship, resort so readily to violent retaliation?

By the end of my tenure, my lion pride had begun hunting and eating people on a semiregular basis. This is not a typical part of lion ecology, and it certainly didn't make it easy to navigate the thin line between care for animals and care for people. Needless to say, it also threw a wrench in my data-collection attempts. Instead of better understanding the relationship between landscape features and lion predation of Maasai livestock, I was forced to respond to incidents of lions attacking humans. This meant doing some unofficial forensics and, most important, speaking to members of the Maasai community about what happened, and attempting to explain what drove the lions to do such a thing.

In one month, three children, two men, and one woman were killed by lions in my pride. Uncovering the details of one particular attack led me to the discovery of a man's head, a shoe,

a bloody spear, and lion prints on the ground. The understand-
able retaliation by the community members led to the deaths of
many wild animals, both guilty and innocent.

I was ethically challenged with suggesting that while the
official wildlife authorities would likely need to euthanize the
guilty lions for public-safety reasons, the community should
continue to support the lion-conservation work that people like
me were undertaking. In doing so, I felt like I was perpetuating
a pro-animal, anti-human approach to conservation.

For the first time in my career, I was faced head-on with questions and doubts about the validity of my work. Prior to this,
my work had always seemed relevant and positively impactful.
Seeing severed human bodies lying in the bush and knowing that
the animals I cared for were to blame broke my heart. I didn't
want to accept that two ancient species—humans and lions—
were unable to live together in harmony.

As livestock herding continued to get pushed farther into lion
habitat because of outside forces, lions were forced to share more
space with people. Preying on cattle was problematic but not un-
heard of. But attacking the people herding the cattle? Everyone
deemed this unnecessarily violent, so the lions were targeted.
From a biological perspective, the lions' behavior was completely
unnatural. The published science, plus my own anecdotal evidence of the lions circling my tent when I was a young woman
in Kenya, had taught me that these animals could be dangerous
but didn't intentionally hunt people to eat.

I wanted to grieve with the community and catch the perpe-
trators, yet I also was compelled to protect the lion prides from
further harm. I had spent much classroom time learning about
human-wildlife conflict in theory. Hell, I was collecting data on
it for my thesis project, which now seemed frivolous when faced

with the actual, lived experiences of the groups involved in the conflict. My privilege smacked me in my face while the issue remained unresolved.

Until this point in my life, justice had always been simple: something was right or wrong, black or white. But shades of gray were appearing. I was learning firsthand about the complexities surrounding the needs of people and the environment, and when conflicts arise, there often isn't an easy solution.

ONE WEEK, Peter and I were doing lion surveys on our own. It was fun with only the two of us because we would hit it early in the morning, and then we'd usually have a lot of extra time after the work was done. Peter loved taking game drives—cruising around, seeing what kinds of animals were out and about, and taking note of them. This wasn't a leisurely activity for Peter, in the way it might be for a tourist or, honestly, for me. He was studying animals all the time.

Peter's big goal was to be a Tarangire National Park ranger. That was another reason I loved the days it was just us: I wanted to support Peter and his preparation for landing the job.

During our game drives, we'd spot zebras, gazelles, impalas, elephants, hippos, and giraffes. Giraffes were everywhere in this part of the landscape. I love this animal for many reasons. The Swahili word for "giraffe"—*twiga*—was one of the first ones I learned. And it took on a second meaning for me because it was the name my four female hut-mates and I gave ourselves during my Kenyan study-abroad experience.

There are so many grazers on the savanna, but giraffes are browsers, and this makes them important. They browse from trees with their black tongues, which can be up to twenty inches long, by wrapping them around tree branches and pulling leaves off—even acacia trees, which are full of thorns. Most animals can't eat acacia leaves, but giraffes have no problem at all.

One day, Peter and I were out driving and found a paved road, which was extremely unexpected in the bush.

Peter, confident and stoic as ever, didn't appear curious about the road, so I figured he was deliberately driving us somewhere. But not knowing where bothered me. I wanted to feel like real teammates, which would require communication and making plans together. From time to time, he'd drive us places, and I wouldn't know our destination until we arrived. We might visit the village to greet an elder or a wildlife watering hole to gaze at the animals while we ate lunch.

Our joint work and excursions were fulfilling, yet when I worked with him, I often felt an uncomfortable power dynamic. I never confronted him about it because I didn't want to offend him or cause our fairly good relationship to sour. That's who I was at the time: a people-pleaser who allowed myself to be treated in ways I didn't like. My worst-case scenario wasn't being mistreated—it was being unliked.

So when I spoke to Peter about our latest adventure, I made sure the tone in my voice sounded innocently curious, not accusatory.

"Hey, where does this road take us?" I asked.

"Oh, I need to drop something off to a friend," Peter said.

Despite being annoyed, I trusted him completely.

We turned onto the smooth paved path and drove up a hill to visit his "friend." At the top of the small hill, the vegetation was cleared out, and a camp with some legitimate infrastructure had been built. As we neared, I saw a meeting area with generators for electricity and a table and couch. On the opposite side, I saw several hammocks hanging between posts that had been nailed into the dirt.

A couple of guys were visible when we pulled into the camp and came up to our truck to greet us. They wore dark-green uniforms and grinned when they saw Peter.

Peter introduced me as Rae, a researcher on his lion project. He finally shared that they were Tarangire National Park rangers. Everything clicked for me then. He was networking.

Although these rangers were serious about their jobs, we'd caught them in a moment of downtime. Everyone was relaxed and happy for a visitor to gossip with, and the energy was loose and casual. For a couple of moments, things felt familiar. Yes, we were in the Tanzanian bush after finishing a lion-tracking expedition, perched on a hill overlooking the Tarangire Ecosystem. But I felt at ease talking, laughing, and hanging out with a group of twentysomething Black men.

The air was full of sounds and speech from all kinds of walkie-talkies and radios, yet no one seemed to be listening to the constant chatter. Peter and I sat on the couches in the cool shade. We stayed for a while, sipping our water and talking about Tarangire and our research.

A different ranger strode over and said something in Swahili so fast that I wasn't able to catch it. Everyone dropped what they were doing, stood up, and ran to their vehicles. Peter told me in English, "Okay, get up right now. We have to go." Something was wrong, and whatever it was, it was serious. We jumped back into our truck.

"What's going on?" I asked with fear in my voice.

"It's poachers. In Tarangire National Park," Peter said, his face grim as he took off behind the rangers.

A feeling swelled up inside me as I realized we might be in danger. A huge theme of my summer had been listening to my primal instincts, and in many ways, that had kept me out of serious trouble. Yet here I was with full knowledge of clear and present danger, and my gut was telling me to flee from it, not drive straight into it. I started panicking, but didn't want to show it outwardly.

"Where are we going?" I yelled over the loud bumping and crashing noises the truck made as we sped along the dirt road. "If there are poachers, where are we supposed to be? Where, where are we going? Are we going back to camp? What are we doing? *Peter!*"

Peter didn't answer me, in the way I'd noticed men often do when I try to get their attention as they're multitasking. I didn't like being ignored, but his adrenaline was spiking as well. As in many other instances that summer, I'd have to rely on myself to get through this adventure.

We followed the park rangers as they drove fast down the dirt road, kicking up a cloud of dust so thick we could barely see through our windshield. It felt like a high-speed chase with cop cars, racing to a crime scene.

We listened to our transistor radio. Harried voices clamored to speak over each other. Peter translated the conversation. "Okay . . . there were three poachers. They've been captured . . . but we are too late." He paused before speaking again. "They killed a giraffe. An adult female."

The magnitude of that news hit us simultaneously. I turned from looking at him to face forward in my seat, gripping the door handle to steady myself as our truck whipped around curves. My brain went into overdrive, trying to comprehend that armed poachers were afoot in our area and that they'd been successful.

And with something as large and majestic as a giraffe. A female, potentially a mother or possibly pregnant. Although I was familiar with the statistics of how frequently poaching happens, I was devastated.

I looked back at Peter and tried to read his face. It appeared mostly stoic, tinged with anger. Tears welled up in my eyes. I took a deep breath and decided not to let them fall.

Soon, we began to slow down as patrol trucks became visible through the bush. Peter shifted our truck into first gear as we carefully turned and began the all-too-familiar, bumpy off-road drive. This time not for research but for emergency response. I could see the giraffe in the distance and realized that I'd never seen a giraffe lying down. Giraffes don't lie down. Because of the weight of their bodies and necks, it takes them a long time to rise even from a seated position, let alone lying down. So giraffes usually rest while standing up. But this female giraffe was lying flat on the red earth, her final resting place.

Peter parked next to one of the ranger patrol trucks. The scene was eerily quiet. The loud engines and yelling men had driven away all the wildlife in the area, including the birds. We were deep in the heart of the Tarangire Ecosystem, the savanna dotted with acacia trees and the occasional boulder outcrop.

I counted six park rangers, dressed in military-green uniforms and government-issued caps, with machine guns slung around their torsos. In the back of one pickup truck sat three young men in tattered black clothing, handcuffed, their heads hanging down. No question they were the people who'd been caught poaching. The tension was palpable. One ranger guarded the truck with the poachers, another was in the driver's seat, and the four others stood next to the giraffe.

I remember her belly the most. It was astonishingly large, and if she were still alive, it would be inflating and deflating with deep breaths. But it was still. I was surprised that the body wasn't bloody. If I'd come across that giraffe on my own, I probably wouldn't have realized it had been shot.

She was beautiful, peaceful and calm, matching the loveliness of the day. Under different circumstances, it would have been amazing to behold her so close.

Peter walked over to the rangers he recognized, and I stayed by his side. During the drive, I'd heard him speak over the transistor radio, explaining that he'd be accompanied by "an American researcher he was training." He must have also described me as a Black woman, because the rangers didn't question who I was. The engine of the truck holding the poachers revved, one of the rangers said, *"Tutaonana"*—Swahili for "See you later"—and they drove off. I never asked where they were going, but I imagined that the drive to the nearest law enforcement office was far.

While various feelings swirled within me, the scientific part of my brain gravitated toward logistics: *Well, what do we do now?*

My mind raced with possibilities. *Maybe they'll take the whole giraffe's body and bring it to a museum,* I thought. *Maybe it can be brought to a university and can be dissected or used for science. Maybe they'll dig a huge hole, and we'll give it a burial.* My speculations were futile—I'd be better off just asking Peter, who stood right next to me. His calm demeanor made me think he was more familiar with this type of situation than I would have expected. He was intently listening to the park rangers speak rapidly in Swahili and didn't notice me trying to make eye contact.

"Je, tutafanya nini a twiga?" I asked in Swahili as I nudged him. *What will we do with the giraffe?* Peter abruptly pulled me away

from the circle of park rangers to speak to me in English. It suddenly seemed like he was embarrassed that I couldn't glean what the next step would be. I began feeling that I had committed yet another cultural faux pas.

"Well, now we're all going to take some," he said.

"Take some of what?"

He stared deep into my eyes as he replied, signaling for me to not question the decision. "We're all going to take some of the giraffe."

A giraffe can weigh thousands of pounds, and this one offered about a ton of muscle tissue. Peter, slightly gentler now with his delivery, explained that the amount of meat sitting in front of us could feed many people for a long time, but we had to act fast. Word had already spread throughout the neighboring Maasai communities, and people were on their way with baskets, bags, and pots. Everyone would line up, hack off some giraffe meat with a machete, and bring it back to their home.

We were in the bush, with no electricity and thus no refrigeration. All of this had to happen while the meat was still fresh and before nightfall, when carnivores would scavenge the carcass. Time was of the essence, and it was only safe for us to stay by the giraffe for a short time.

While my mind had been racing with potential next steps moments before, this idea hadn't dawned on me. Of course, there was an opportunity to use the whole animal. What, were we going to let it sit there and rot? Absolutely not. And of course, this community of indigenous people, who had shared the land with wildlife for millennia, already had a protocol for when an animal is illegally killed.

A glimmer of hope grew within me. I didn't have to be completely powerless in this tragic situation—I could be active in a solution. Optimism and action are parts of my core essence, and in that moment, I blossomed into the Rae who learns, understands, and helps. I ran over to our truck and grabbed two big machetes. Before I knew it, Peter raised his arm high, almost like he was wielding an axe, and brought it down hard, hacking into the hip area and cutting meat from the dead giraffe's body. I watched as more men joined in, and part of me was eager to get it done.

Peter handed me big pieces of giraffe meat, some still wrapped in skin bearing the unmistakable pattern.

I estimated that each piece was roughly twenty pounds as I hauled giraffe hunks to the back of our trunk over and over again. I was getting into the rhythm when I glanced up, my eyes spotting something moving on the horizon about half a mile away. In the grasslands, it's easy to see far, and after having spent so much time training my eyes to see lion movement in the distance, I could spot anything.

A single-file line of Maasai women appeared, beautiful cloths draped over their bodies, slowly approaching our giraffe-poaching site. Brilliant reds and blues and whites against dark-brown skin created a rainbow of vibrant hope. Around fifty women came, many of them carrying big, empty baskets on their heads.

It was powerful to see women I believe to be the foundation of their communities walking, for what must have been several miles, to the site of a dead giraffe. The giraffe's unnatural death— a violent crime—was extremely upsetting to Peter, the other park rangers, and myself. My conservation-science training drilled into me that any and all poaching was bad. Yet here were the Maasai women, who would now be able to bring several days' worth of essential protein back to their homes, for their children, partners, families, and community.

This is what's missing from the conversations in my classrooms, I thought. *The nuance, the duality, the fact that the death of one can transition right into life for another.* I also thought about the way the inequalities of incarceration were bundled up in this situation. Three young Black African men had been arrested and would likely go to prison, ruining their lives and potentially the lives of their families. For all anyone knew, their choice to take on a physically dangerous and legally risky job opportunity had been made out of desperation rather than greed or bloodthirst. And would incarcerating these men truly be an important step toward reducing wildlife poaching in this part of Tanzania? Or would these foot soldiers of the illegal wildlife trade be replaced by whichever kingpin was managing this crime ring from a place of protection?

Just as I'd seen back home in America, I was beginning to realize that perhaps the issue at hand wasn't wildlife crime but the resource needs of Black communities.

"Okay, Rae, I think we have enough for this car!" Peter's voice jolted me out of my thought spiral and back to the task at hand. I'd been on autopilot, catching cuts of meat and tossing them into the truck, not realizing how much we'd loaded up. I looked at my palms, stained dark red with giraffe blood. I instinctively lifted them to my nose to smell and inhaled a sharp odor I can only describe as liquid warmth.

Wildlife poaching is a tragedy—full stop. I believe these animals have inherent value and should live beautiful, full, safe lives. And scientifically, many of these animals are critical ecosystem engineers, without whom our environments would collapse. I also believe that the illegal wildlife trade—second in magnitude only to the illegal drug trade—needs to be ended, and soon. At the same time, based on my lived experiences, community members' input, and my own research, I believe it's important to expand our understanding of what poaching is and isn't, and especially how we got to this place.

Giraffes, once in the multimillions across sub-Saharan Africa, drastically declined in population during European colonial rule, largely as a result of the sport hunting and land development driven by economically powerful white people. So today, with fewer giraffes than ever and highly racialized economic inequality across Africa, one person's life can be snuffed out because of killing one giraffe. My science background had trained me to only see the wildlife problem and stay blind to the human problem. Yet there I was, hands soaked, waiting to greet women from the nearby villages as we turned a setback into an opportunity.

This moment reminded me that humans are a part of the environment. Human life, human dignity, and human livelihoods play a necessary, inseparable role in the health and functioning of ecosystems. While this may seem obvious, it's a difficult concept to keep at the top of your mind when your whole career is oriented toward saving nonhuman animals.

At the same time that I was working so hard to understand lions, to figure out how to best protect them, there was an entire community of people living in these ecosystems. We shouldn't consider any lion project or environmental project complete if the human communities in those places have unmet needs. In the conservation community, we need to prepare ourselves for the infrequent times we must prioritize human well-being, potentially over an environmental goal. If any compromise needs to be made in the future, it might need to be made for the benefit of human dignity and well-being. And that should be seen as a win, not a loss.

By the time Peter and I returned to our field station, word of the adventure had preceded us. As had a large hunk of giraffe meat, which was stewing in our cauldron, along with the handful of onions, tomatoes, and spices that we had. Peter and I hadn't taken any meat for ourselves, knowing that every available piece should go toward feeding a person with higher nutritional needs than our fairly privileged research team. Yet one of the truck drivers had delivered us some anyway, a gesture of appreciation that represented the deeply held community value of sharing.

The meat tasted terrible to me, which was saying a lot. I'd already spent months eating some extremely unfamiliar things yet enjoying them. The meat was tough and rich in flavor, very gamy. After spending so much time with the deceased giraffe's body and handling so much of its flesh, I could no longer separate the smell of the giraffe from the taste of the stew. Not to mention that stew needs to simmer for hours, to tenderize the meat and release all the flavors, and this one hadn't been cooking nearly long enough to get those benefits. But out of respect for my team, who cleaned their plates—and even more, out of respect for the gorgeous wild animal that had lost her life that day—I ate my stew down to the last drop.

AFTER THAT SUMMER, I returned to New Haven with a deeper understanding of the interrelationships between animals and the humans with whom they coexist. My time as a Black American woman living within a Maasai community was irreplaceable. Although it was a time riddled with misunderstandings, close calls, and cultural faux pas that made me long for the familiar culture back home, in the darkest and most confusing times, and at the moments when unexpected death and tragedy came knocking, the Maasai community was there to hold my grief and share my pain. Thousands of miles away from America, in a land where centuries-old tribal lifestyles are still practiced daily, live our brothers and sisters concerned with Black lives and liberation, and a lion named Michael Jackson. We are welcome there anytime.

Chapter 5

During the fall semester of my second year at Yale, Dr. Shahid Naeem, a renowned wildlife ecologist, was invited to speak at a major seminar on campus. My co-advisor, Dr. Dave Skelly—a superstar ecologist at Yale who went on to become the director of the Peabody Museum of Natural History—knew that I wanted to pursue a PhD in wildlife ecology and recommended that I attend Dr. Naeem's lecture because many faculty members from several high-ranking universities would be there. Dr. Skelly saw this as a great networking opportunity for me.

Dr. Naeem was the chair of the Ecology, Evolution, and Environmental Biology Department (E3B for short) at Columbia University in New York City, and many of his E3B colleagues came up to Yale to support his speaking engagement. I attended the talk—most of which went way over my head—and I found Dr. Skelly afterward. My timing was perfect, as he was in the midst of greeting some of the Columbia faculty.

We made eye contact, and he gestured toward me and announced to the small group around him, "I'd love for you all to meet Rae—she's one of our most serious wildlife ecology students, and she's just come back from a summer tracking lions in Tanzania." That elicited a positive reaction from the group, and Dr. Skelly added, "She'll be applying for PhD programs soon, and I can vouch that she would make a strong candidate for any of your labs."

I hadn't expected Dr. Skelly to personally introduce me to some of the most influential ecology professors on the world's stage—and I definitely didn't expect him to drop the hint that they should consider me for their PhD programs.

To him, this exchange was merely a couple of sentences in the middle of an enjoyable social event. But for me, it may have been the exact point of inflection where my trajectory solidified.

Despite Dr. Skelly hyping me up, I was still shy and attempted to slip away from the event as it was winding down.

"Excuse me, is it Rae?" a voice asked, and I turned around. In front of me stood a fairly petite white woman wearing a scarf that looked like it came from somewhere in East Africa. She had glasses and faint wrinkles that made it hard to gauge how old she was—certainly middle-aged, but there was a youthfulness to her energy and tone of voice.

I introduced myself to her and learned that her name was Eleanor Sterling. She asked how serious I was about PhD programs and what I wanted to study, and she told me that although she was currently the chief conservation scientist at the American Museum of Natural History, she was an adjunct faculty member in E3B and able to take graduate students. She also emphasized that although the department was grounded in ecology, and particularly the philosophies around ecological theory, some professors, like her, were conservation practitioners—their projects had the sole purpose of solving wildlife conservation challenges.

Our conversation encouraged me to apply to Columbia, in addition to three other programs I was already considering. And I got rejected by all of them.

The situation was disheartening but also showed my complete lack of understanding about PhD programs. I thought that one degree built upon another: you got a master's degree, and then you got your PhD. What I didn't realize was that a master's degree was rooted in practice, but a doctorate is rooted in theory. My admission applications weren't as sophisticated as they could have been, and I would've been better prepared to demonstrate my aptitude for a PhD program if I'd had an opportunity to ask the right people the right questions.

In Ivy League circles, many people assumed that if you were already part of that world, you knew how to operate within it. But I felt like an interloper in this elite group and didn't have access to the same generational knowledge as my peers. I didn't know what came next or how to achieve it. I just knew that I wanted it. And because of my ignorance—and society's assumptions—I almost missed the opportunity to progress to the next level of my career.

ONLY A FEW WEEKS after I'd received my Columbia rejection notice in the spring of 2010, I attended the inaugural Student Conference for Conservation Science in Cambridge, UK. I was there to present my master's thesis work, a notable achievement, but disappointment overshadowed my feelings of pride. I was mortified when I saw Eleanor there and hid from her for the first two days of the conference. She finally found me and told me that I was rejected because my test scores were too low. However, she knew how serious I was. Between that and Dr. Skelly's strong recommendation, she was determined to see if she could fight for me to get in and for her to be my advisor.

And that's exactly what she did. I was eventually admitted to the E3B program with full funding for all five years.

What I didn't know at the time—and wouldn't discover until I was well into my first year in the program—was that I would be Eleanor's last PhD student. Due to her job duties with AMNH and the way the E3B department was restructuring professorships, Eleanor would no longer be able to take on PhD candidates. Apparently, she'd fought to even be able to admit me into her lab and had promised the department that I'd be her final student. This revelation made me feel both honored and grateful, but I also felt an immense amount of pressure. Eleanor was risking her own reputation by admitting me and taking me under her wing, and the stakes felt high. After all she'd done for me, I couldn't afford to be anything less than excellent.

AFTER FINISHING MY master's degree in 2010, I traded the hallowed halls of Yale for the concrete jungle of New York City. I knew the PhD program would be extraordinarily rigorous, so I decided to move to the city in June, to have at least one summer without school demands weighing me down. At age twenty-four, I was ready to feel the city's energy and spend some time exploring the place that would be my home base and launching pad while I attended Columbia.

August rolled in, and the summer fun ended as I immersed myself in school, working and studying pretty much all day, every day. The decision to get my own apartment turned out to be one of the best I'd ever made. When I got home, I had my own space to be focused on my studies and to take care of myself. I didn't feel compelled to people-please, I didn't have to worry about dealing with anybody else's drama. My apartment became my personal sanctuary.

At some point that fall, the one girlfriend I had in New York called me.

"Rae, what's going on? I haven't seen you in weeks. You never go out anymore," she lamented.

"I know—I've been really busy with school. I have to study and keep my grades up, so I can't be going out all the time," I told her.

"Well, how do you expect to ever find a husband if you're not going out?" she asked.

The word *husband* sent a bolt of anxiety through me. Somehow, I'd internalized the idea that twenty-five was the optimal age to begin the stereotypical relationship trajectory: dating, engagement, marriage, children. Having recently turned twenty-five, her words triggered a sense of urgency.

I hadn't been in a serious relationship since I'd lived in Washington, DC. But my thinking about dating wasn't limited to the idea of finding my future husband—I was ready to love someone, and to be loved by someone. I was pouring myself into my studies and professional work, which were good for the planet and society. I had plenty of outward passion, but I wanted something interpersonal. I felt like I had so much to offer and so much to give.

"Okay, the next legitimate event you have, I will go with you," I agreed.

A couple of weekends later, she called me. "I found this happy hour event, and a bunch of eligible Black men will be there," she said.

The fundraiser was for a nonprofit organization that supported underserved Black youth. It was being held at a popular club, one we'd often go to at night. I was relieved that it was being held during the early evening, so I could have a night out but still get home at a reasonable time.

My friend insisted that we get dolled up. We put on tight dresses and heels and did our makeup and hair. Then we took the subway to the club in Midtown Manhattan where the event was taking place.

When we walked in, we saw that everyone in attendance, men and women, were dressed in business casual attire and kind of conservatively. And there my girlfriend and I were, decked out like we were going to the club. We looked anything but professional, our youthful inexperience and ultimate interests quite obvious.

I felt uncomfortable and self-conscious. But then I figured, *You know what? I'm here. I might as well make the best of it.* I didn't mind male attention, so I went for it.

My friend had been right about the place being full of young, professional Black men. Most of them seemed to be about ten years older than us, around thirty-five years old. I wasn't used to dating outside of my age group, but maybe a ten-year age gap wasn't so bad.

Although I wasn't much of a drinker, I figured I might as well start my mingling at the bar. A group of guys who seemed like they must be old friends were ordering drinks, and their relaxed, fun energy appealed to me. By then I had abandoned the idea of finding my husband at this event, but at least I could meet some good people.

I squeezed my way to the front of the group, introducing myself and shaking hands along the way. One of the more gregarious men introduced himself as Oba. Oba had a strong, muscular build and dark-brown skin. He was wearing a dark-blue suit with a white dress shirt underneath, and although he'd entered the event with a tie on, he'd loosened it and pulled it off as he sunk into conversation with his friends and colleagues. He had an endearing laugh that made his eyes squint and a cute, youthful smile. We ended up in conversation and got along immediately. When I told him I'd recently moved to New York, he asked where I lived.

"Harlem," I said

"Oh, I live in Harlem too," he said.

"Where?" I asked him.

"On 122nd Street and Frederick Douglass."

"I'm on 115th and Frederick Douglass! We're basically neighbors." I marveled that I'd met someone who only lived a few blocks away from me.

"I was born and raised in Harlem. So if you ever have any questions or just need a neighbor or something, let me know. I'm around."

He wasn't being flirtatious or suggestive—he was offering friendly support to a new girl in town. In spite of his offer, we didn't exchange phone numbers.

I didn't stay long at the event, as I couldn't shake the desperation I had arrived with. It had been years since I'd been in a real relationship, and since I was now focused on establishing a long-term commitment, I wanted to attract a partner through a spirit of abundance, not lack.

After chatting with only a couple of guys, I abruptly told my friend I was heading out and hailed a cab. When I arrived at my building, I walked up the five flights of stairs to my apartment, kicked off my heels, wiped off my makeup, pulled on my sweatpants, and reopened my textbooks. Maybe I'd meet someone special that winter, or maybe I wouldn't. Regardless, I needed to show up as my authentic self—even if that self was a tired, overworked graduate student.

About a week later, I was checking Facebook and saw that I had a message from Oba:

Hey, Rae! We didn't exchange numbers at the event, so I'm hoping this Facebook message gets to you. It was nice to meet you, and if you ever want to link up in Harlem, I'd be happy to show you around or show you some of my favorite spots.

If this guy had gone through the trouble to track me down on Facebook and send me this message, I figured he was trying to ask me out on a date.

I found his respectful approach and the absence of pressure attractive. I responded and said that I'd love to get together, but I'd soon be cramming for final exams and wouldn't be going out for a while. Maybe another time? I also tried to make it clear that my reasons were legitimate and that I wasn't blowing him off.

He wrote back and said he totally respected that, and that if I ever needed a study break I should let him know.

Two more weeks passed, and one day, deep in the mire of studying, I acknowledged I was miserable and needed to rest my brain. I recalled Oba's offer and sent him a Facebook message that included my phone number. I told him I was ready for that study break and asked if he was free that Saturday. I was giving myself one hour, 5:00 to 6:00 p.m. I said I'd meet him at the beer garden across the street from my apartment.

When we met up, we realized neither of us particularly liked beer, which actually put me at ease. After having spent so much time surrounded by the ever-present drinking culture in academia, it was refreshing to meet someone who didn't feel that consuming alcohol was essential for every social event. Oba was already showing me that he was different and more mature than the guys I was used to. Typical of a first date, our conversation was awkward at times, but overall, the dialogue was pleasant and flowed easily. He asked me a lot of questions, which kept the conversation upbeat and interesting, and it was a novelty that he didn't spend the whole hour talking about himself.

Among the many questions, Oba asked me who my favorite musical artist of all time was. I initially said Michael Jackson, but then I reconsidered.

"Actually, you know what? It might be Prince," I said.

"Whoa, that's amazing! Prince is my favorite musical artist of all time too."

"Wow, really?"

"Hey, this might sound crazy, and I've never done anything like this before. But I have two tickets to see Prince next week at Madison Square Garden. I haven't been sure who to take. Would you like to go with me?"

"Are you kidding me? Of course I'd love to see Prince at Madison Square Garden!" Prince was touring for the first time in about a decade, and this was a fantastic opportunity to see an artist I'd loved and admired my entire life.

This invitation was the highlight of what ended up being our first date.

At the end of the hour, we hugged and said goodbye. I was excited to have something to look forward to in the midst of all my studying—both the concert and a second date with Oba.

Oba and I stayed in regular contact. Before the concert, he invited me on a proper dinner date. Once again, I enjoyed his company and the conversation.

He was thirty-five years old, ten years my senior, and I'd never dated someone with such a significant age gap. Oba seemed so different from any other guy I'd dated. He had a stable career and strong relationships with his friends and family. He seemed like a genuinely happy and satisfied guy, secure in who he was—confident but not arrogant. The only thing missing in his life was a relationship.

Did he make my heart race and adrenaline pump through my veins? Not on those first two dates. But I didn't feel anything negative or questionable either, and I was drawn to the idea of having a partner who was strong, steady, and reliable. And those were the vibes I got from Oba.

We went to the Prince concert, and as soon as we arrived at our seats, Oba sweetly expressed that he hadn't wanted to come off too strong on our first couple of dates, but he'd very much wanted to kiss me.

"In case it isn't crystal clear, I really like you and would love to get to know you better and take you on more dates," he added.

This clarity astonished me, and I found it incredibly attractive. After a decade of dating guys who kept me guessing about their level of interest, here was a grown man being forthright with his intentions and leaving the ball in my court.

"I'd like to see you more too," I offered flirtatiously. Our exchange was sealed with a kiss, as the Madison Square Garden lights dimmed, and the crowd began to cheer for Prince. We had no way of knowing that in a few years, this magnificent, brilliant performer, someone who was part of a major milestone in our lives, would no longer be with us.

One of the highlights of the evening was when Misty Copeland performed with Prince. It was an incredible moment, with a famous Black musician and the first Black prima ballerina for the American Ballet Theatre sharing the stage at one of the world's most famous venues.

However, at the time, I had no idea who Misty Copeland was. I asked Oba if he knew who this graceful, powerful dancer was.

"That's Misty Copeland," he told me. "I know her—one of my friends is dating her, and we all hang out. She's a great person in addition to being a talented artist."

He texted his friend from our seats, expressing how cool it was to see Misty onstage. A year later, when Oba and I were in a long-term relationship, we'd double-date with Misty and her boyfriend. And years later, we'd be guests at their New York wedding celebration.

After that night at the Prince concert, Oba and I officially began dating. He was so different from many of the previous guys I'd dated. He was nice and drama-free. And he was a great boyfriend. Interestingly, our age difference was most highlighted by where we chose to invest our energy. I was on an ambitious years-long climb to solidify myself as a scientist and wildlife expert, and despite having had years of adventures in East Africa, I believed that many of my most profound career moments were still ahead of me. Over the previous decade, Oba had been through his own career ascent and had landed in a solid, notable position that he was comfortable with. He wanted to settle into the stability he had earned, and rightfully so.

The month after the Prince concert, I was talking to my mom about him, and I told her, "Mommy, I don't know if he's the one. Or if I'm just making better decisions about who I date."

Within two months, Oba and I had the conversation: Are we in a relationship? Yes. Does this feel good for us? Yes. Is it a serious relationship? A serious monogamous relationship? Yes and yes. Having that commitment was so appealing, and I took his willingness to commit as a signal that we were meant to be together.

Less than a year after we started dating, we moved in together. My lease was ending in June 2011, and though we didn't necessarily feel the time was right for us to live together, the desire to save money in one of the world's most expensive real estate markets won the day. I was somewhat reticent about giving up the personal, private space I'd come to cherish. It also would be a big adjustment for Oba, who'd never lived with a girlfriend.

Thankfully, we didn't have to navigate the challenges of cohabitation right away because soon after I moved in, I left for my first summer of fieldwork for my PhD research. In the coming years, I'd be spending a lot of time traveling to and from Lake Tahoe, which allowed me to maintain some level of independence.

I ignored the red flags waving early on because I desperately wanted to be part of a relationship that worked. I'd dated few men who wanted to commit to me (or to anyone, for that matter), and here was Oba, who was willing to commit right away. Additionally, the red flags were easy to ignore because they weren't warning me about trouble that Oba would cause in my life. Instead, they were more subtle, representing a general lack of compatibility. We were both good people interested in love and growth and stability. So without much experience or deep thought, I dove right in.

IN ADDITION TO navigating the joys and challenges of my new relationship with Oba, the Columbia doctoral program was rigorous, technical, competitive, and unwelcoming. For the entirety of my doctoral studies, I felt out of place. I was no longer sleeping under the savanna stars but staying up late into the night, the darkness illuminated not by galaxies in the open sky but by the glowing blue light of my computer screen. I spent hours at a desk reading and writing theoretical work and scholarly articles.

The first semester, I received a call from Eleanor on my cell phone.

"Hello?" I said, marveling that this was the first time a professor had ever called me, as opposed to their usual emails.

"Rae, you're failing your ecology class," she said without any preamble.

"What? *Failing?!*" My disbelief reverberated through the phone.

"Well, right now you have a B minus," she said.

"Oh, I know I have a B minus," I replied as I relaxed somewhat. Eleanor had scared me there for a moment.

"In Columbia's PhD program, you must get a B or higher to be considered passing," she asserted.

My momentary sense of relief had vanished, and it registered in my brain that if I didn't maintain passing grades in all my classes, I'd be kicked out of the program.

My confidence crashed, my grades suffered, and my motivation stalled. Knowing that my clock was ticking, I searched high and low for opportunities to return to the field, to the familiar. So when the chance to initiate research in Lake Tahoe presented itself, the anxieties of the classroom fell away at the thought of adventure.

Chapter 6

Nestled in the breaks of the Sierra Nevada lies the cerulean depths of Lake Tahoe. Rocky shores border the lake as snow-capped mountains disrupt the water's infinite stretch to the horizon. Fir trees and stately pines flank the shores and provide shelter for the region's wildlife: yellow-bellied marmots, mountain lions, American martens, and, of course, the largest of the Sierra carnivores—black bears.

I had been studying black bears since I started my PhD in the fall of 2010. And for that first year, while I was meant to become an expert on the subject, I still hadn't seen a bear in real life. I felt like such a fraud—rigorously starting my research career on an animal I'd never encountered.

My new project would take me back to the same region where I'd first explored the outdoors as a schoolgirl on my Yosemite National Park class trip. I would be researching a small black bear population in the Lake Tahoe Basin, on the California-Nevada border, in the middle of the Sierra Nevada mountain range. From 2011 to 2013, I'd swing like a pendulum between my research project in Tahoe and my home base in NYC.

Early on in our research, as a rite of passage, my research collaborators, Dr. Jon Beckmann and Carl Lackey, took me on a long driving tour of all the different habitats bears use in the Lake Tahoe ecosystem. For the first time, I saw the mountains, forests, lakes, and deserts of my birth state and the bordering state of Nevada, all within a few miles of each other. That same day, they drove me to a secluded area of the forest and patiently taught me how to shoot a tranquilizer gun, a tool I'd use throughout my many years of work with black bears.

The timing couldn't have been more ideal, because the next day we captured what we all called "Rae's First Bear." Its fur wasn't black, but light brown, which is typical for western bears. As I'd soon learn, North American black bears come in various shades, even stark white, and their coats often correspond to their native regions.

I learned how to process the bear, which wasn't much different from how I'd process lions in my earlier years studying African wildlife: weigh and measure the animal, check its temperature (rectally—you haven't lived until you've done this), comb through its fur to look for ectoparasites, take hair and blood samples, and give it an ear tag and a GPS collar so we could track its movements and learn about its ecology.

Although I always want to appear to be a cool, collected, well-seasoned field biologist, it was impossible for me to contain my excitement during this first experience catching and tagging a black bear. As soon as it was tranquilized, I pulled out my smartphone and texted pictures of me and the giant animal to my friends and family. The next day was another bear, and the day after that, another. The work energized me, and I was awestruck by how naturally black bears fit into this ecosystem that housed both people and wildlife.

Yet what seemed a "natural fit" to me didn't necessarily reflect the experiences of many people I met during my first days in Tahoe. These people, having built homes and livelihoods in bear country, often perceived bears as a nuisance and occasionally as a threat, either to their safety or to their prosperity. This tension between humans and bears on this shared landscape struck me as a conflict in need of mitigation, a problem I was determined to use science to solve. What came out of my commitment, however, was less of an ability to make sweeping recommendations to eradicate human-bear conflict in Tahoe and more of a deep understanding of the ways human lives and values are distinctly intertwined with bears.

Days later, we captured an adult female black bear with her two "cubs of the year," about six months old. They wrestled and played with each other while we processed their mother, placing a GPS collar around her neck that would allow us to track her movements into the winter as she made a den for herself and her cubs. Once we had finished, we hid in a nearby bush to monitor the cubs' safety while the mother slowly arose from her sedated state.

It is rare for people to be able to observe the behavior of a mother bear with her cubs, as this is often a dangerous scenario. Watching the playfulness of the little ones, as well as their seeming sense of relief when their mother began to stir again, connected me to these animals in a way I'll never forget. Upon waking, the mother's first order of business was to nurse her cubs, an instinct familiar to all mammals. Reunited and well fed, the troupe of three then calmly walked back into the depths of the forest.

I was hooked on this work for life.

I RETURNED TO NEW YORK at the end of summer in 2011 and dove into my second year of coursework. I needed a few more ecology and conservation classes to fulfill the requirements for me to advance to PhD candidacy. The classes were more manageable, and I was able to participate more actively in the lab group meetings that my advisors, Eleanor Sterling and Joshua Ginsberg, ran for students who were excited about conservation. Some sessions offered mentorship, teaching us safe fieldwork practices, and others were group research efforts examining the different measures used to protect animals on the endangered species list. I passed my classes that year with flying colors, finally finding a groove with school, my social life, and my relationship with Oba, which had become very serious, very quickly.

By summer 2012, I was headed out to Tahoe again, to work with Carl and Jon and continue our bear research. With some experience under my belt from my first summer, combined with all the academic expertise I'd gained at Columbia, I no longer felt like the new girl on the project, and Carl certainly treated me like someone who belonged. He offered me extra responsibilities and even sent me out on my own a few times to collect data.

Having some experience didn't necessarily make the work easier. We were still on a mission to place GPS tracking collars on what we called "wild bears"—the ones that lived deep in the wilderness and thus had never come into contact with humans. I was camping in the backcountry of Nevada, sometimes with only Carl's Karelian Bear Dog, Rooster, to keep me company. Days—and sometimes a full week—would go by without a single bear taking the bait, leaving both my traps and data sheets empty.

I knew this was an important part of the process and a particular lesson in science leadership, not to mention patience. But a certain type of stress came with navigating two types of extreme environments at once. While I was experiencing the stillness and simplicity of being immersed in the arid Nevada backcountry, I was simultaneously expected to be a constantly productive academic, lest I fall too far behind the other PhD students in my cohort. There was a difference between coming to Carl with no collared bear to report and returning to Columbia without a significant amount of data collected.

During my time in Tahoe, I began to feel the tension between on-the-ground work and the publish-or-perish pressures of the PhD process. Finally, I was feeling a sense of belonging in the American wilderness, and the urgencies of wildlife conservation work seemed at odds with the duties related to peer review and conference presentations. I began to see that mainstream academia wasn't where I wanted to be—I desired more purpose, more action.

Returning to New York after my summer fieldwork, I was met by the usual culture shock of embracing the other side of my double life, as well as a huge surprise: Oba proposed. He'd asked both of my parents for their blessing and planned a private proposal with care. I was stunned when he dropped to one knee. We had talked about marriage, and we were both clear that we were moving in that direction, but we'd only been dating for a year and a half. As he waited for his answer, while slipping a beautiful custom-made ring on my finger, I thought, *Yeah, this is fine. Who wouldn't want this? I wish this proposal weren't happening so early, but we're going to get married eventually, one way or another.* I said yes.

As 2012 transitioned into 2013, that spring found me immersed in two serious projects. First, I was studying for my oral exams, which my department allowed students to focus on exclusively for several months to prepare. Second, I was planning our wedding. Both events were scheduled to occur in April 2013. The first few months of the year were marked by stress and self-doubt. I worried about how both of these events would affect my future. Passing or failing my oral exams would determine whether my department would allow me to continue on or if they'd kick me out. And the wedding that Oba and I wanted to have was well beyond our means. We had to figure out how to have a wedding on a budget, pay for the majority of the expenses ourselves without going into excessive debt, and impress our dozens of friends, who expected the party of a lifetime. The potential failure of both events preoccupied my mind.

Neither event was a success, yet neither was technically a failure. I didn't pass my oral exams, even after sitting across the table from my committee for more than two hours, having answered more questions right than wrong. After a long deliberation, they decided that some of my knowledge was underdeveloped, and I'd need to write a detailed paper on a few evolutionary biology subjects and have it approved before advancing to candidacy. I was disappointed and embarrassed. I hung my head as I left my examination, walking past my classmates who waited outside the door to congratulate me. Writing the paper, turning it in, and eventually being granted permission to move upward in the program didn't feel triumphant. Passing this way didn't feel like a win.

Similarly, my wedding day had its own embarrassing moments. The ceremony, which was held in my mother's backyard in Hampton, Virginia, was lovely, and we were surrounded by family and friends. However, our reception started later than planned, and the caterer ran out of food before more than half of our guests had been served. Dozens of our hungry, tipsy guests hugged and kissed us goodbye, as they left our wedding early to go find something to eat. Oba and I went back to our bed-and-breakfast married and generally happy but tinged with shame. We'd planned a short three-day honeymoon in Montreal because, with my newly bestowed PhD candidacy, I needed to get back to Tahoe to focus on data collection.

By summer 2013, I was itching to get back in the field. After the stress of my exams and the wedding, I longed to be free from the expectations that hung over me in New York. And by the time I rejoined my colleagues, I felt like an expert in the practice. Every summer since the initial trip with Jon, I'd trapped anywhere from ten to fifteen bears. By the time I arrived in Tahoe that summer, I was comfortable in my abilities to perform in the wild.

My days in Nevada were overwhelmingly filled with really hard work. It was critical to have a solid plan at the top of every day because there was always so much to do. Most of the time, I was doing fieldwork: collecting data, catching bears, and tracking their patterns. Equally important were the days I stayed home in front of my computer and did hours of data analysis, mastering high-level statistical techniques that were brand-new to me.

I was collecting data that had never been collected before. Certainly, other people had put GPS collars on bears, but this had never been done with *these* bears in *this* ecosystem, under these conditions of human development and climate change. I alone was in control of this tiny aspect of scientific discovery.

At the time—and even more so today—climate change was driving tremendous droughts in the West, leading to significant increases in forest fires. My research focused on how bears reacted to environmental challenges such as forest fires, drought, and highways being built in the middle of their habitat. None of these factors had ever been researched in the Lake Tahoe region. The conclusions I was making about bear ecology and the types of landscape changes humans bring about would be crucial to other conservation projects all over the world, in places where human development continues to encroach upon the natural

habitats of wildlife. Our work was meant to protect these animals, to protect their habitats and facilitate the balancing of what
was becoming an unequally yoked ecosystem.

In 2011, the state of Nevada had legalized a black bear hunting season in the Lake Tahoe Basin. Data from state biologists,
including those with whom I had worked closely, demonstrated
that the black bear population in Nevada was growing rapidly
enough that a few bears could be harvested each year without
negative impact to the population. For many people in Nevada,
both trophy and subsistence hunting are sources of major cultural
pride and are popular forms of recreation. Elk, deer, sage grouse,
and even mountain lion hunting were legal in the state, and the
legalization of black bear hunting was something many locals
were eager for.

On the flip side, many Nevada residents were outraged over
the decision and organized into powerful advocacy groups to
protest, litigate, and voice their objections to the hunt. Some
argued from the animal-welfare platform, while others suggested that the data on black bears was not yet sound enough
to make definitive claims about a hunt. Misconceptions about
the types of people who were pro- or anti-hunt were rampant,
and I often heard accusations that the anti-hunt groups were
the wealthy transplants from Silicon Valley who were trying to
change Nevada's cowboy culture—that they were people who
didn't know anything about big-game hunting. I also heard views
that the wildlife authorities didn't want bears in the state and
therefore manipulated the data to allow for them to once again
be hunted to extinction. As with most volatile political issues,
the general message was lost. In actuality, most people wanted
an intact and functional ecosystem, including a healthy wildlife
community.

*   *   *

FOR BLACK BEARS TODAY, humans are a huge threat. In North America, human-induced causes kill more black bears than any other cause. Usually it's direct sources—hunting, public-safety euthanizing, illegal poaching, vehicle collisions, etc. But what is less talked about are the indirect ways that humans are influencing black bear life and death.

In the Lake Tahoe region, every so often my colleagues and I got reports about mother black bears who abandon the newborn cubs in their dens because of snowmobilers who cause a ruckus and disrupt their slumber. If the cubs aren't found in time, they die. Examples abound, but nothing was more surprising and memorable to me than when Carl and I were summoned to solve the mysterious death of a specific black bear.

Carl called me midmorning, which was unusual. Normally, if he called me at all, it was at dawn to tell me about the bear conflicts that had occurred overnight and that we needed to get up and address them right away. This day, it was almost 10:30 a.m., and I had already sat down to a day of data analysis: organizing the GPS points of collared bears so I could start creating statistical models to predict their movement and behavior.

On that particular day, I was beginning to feel the inevitable desire for a change of pace. Fieldwork was and continues to be my favorite part of wildlife ecology, but after a couple of months, it's common for one's mind to drift to the other extreme—a desire to return from the field, to be back in the comfort of home among friends and loved ones and, in my case, reunited with my city wardrobe.

Carl explained that a call had come in to the Nevada Department of Wildlife dispatch radio about a dead bear high up in the Sierras, not far from a popular hiking trail. The caller couldn't provide an exact location, so Carl warned me that we might be in for a long search to find the animal.

I had never considered that the wildlife authorities needed to be summoned to find dead bears in the forest. To me, it seemed like a bear dying in the forest would be a normal and natural occurrence. Perhaps other wildlife would scavenge its body, or it would decompose and return vital nutrients to the soil—the ultimate circle of life.

As I pulled off the main highway and onto a dirt road with signs marked "Authorized Personnel Only," I spotted Carl's white truck with the "Bear Aware" slogan emblazoned on the side. His dogs were probably in the back of the truck, and I hoped he'd let them use their keen sense of smell to guide us toward the dead bear. Carl is one of the few wildlife managers in the world privileged to raise and train Karelian Bear Dogs, a special breed from northwestern Europe originally used to track brown bears for hunting, now being used in North America to humanely aid in teaching "problem" bears to stay away from human areas after they've been released. The dogs were named Striker and Rooster, both females, and both excellent at their jobs, though they were used sparingly to conserve their energy for essential cases.

Carl hopped out of his truck to run over to my car window and explain the process. We were technically on the Nevada side of the mountains, but it was possible we'd cross into California while on our trek—which might catch the attention of the California Department of Fish and Wildlife. He recommended that I park my car at the entrance to the dirt road and drive the rest of the way with him, since he was official and well-known in the wildlife-biology community.

I couldn't help but wonder if he was looking out for me because my identity as a twentysomething Black woman made me stand out almost everywhere in that part of Nevada, and he didn't want me to encounter any hostility. Numerous times during my field season, I'd considered telling Carl about some of the uncomfortable interactions I'd experienced, although few of them were hostile.

Following Carl's suggestion that I ride with him, I parked my Land Rover and spent a few rushed minutes shuffling around for the supplies I needed. It was October, and the western fall was in full force, which meant drastically different weather up in the mountains than down in the Carson Valley, only a few miles east. Although the sun was out, I wrapped my well-worn black fleece around my waist, remembering the Girl Scout motto from my youth to always "be prepared."

Carl's truck smelled mildly of black coffee and the faint remnants of the cigar he'd smoked the day before, which reminded me of my first ride through the Sierras with Carl in the summer of 2011, when I was twenty-five years old and had just flown in from Manhattan. He asked me if I smoked cigars, and I recall thinking that nobody had ever asked me that question before.

"No, but smoke doesn't bother me," I lied, silently calculating how long my asthmatic lungs would last, trapped in a truck filled with cigar smoke. I soon learned that Carl smoked cigars in celebration and found reasons to celebrate as often as possible—an ethos I greatly appreciate.

As the memory of my first bear flashed through my mind, elicited by the lingering cigar aroma, we climbed into Carl's truck. He explained that a recreational hiker from California was walking his dog and exploring this part of the Tahoe Basin. The hiker claimed he and his dog stayed on the trail the whole time, but he spied what looked like a bear in a nearby stream. Even though it wasn't moving, he obeyed the bear-safety rules he'd grown up with and didn't approach the animal. He finished his hike and called NDOW to report the sighting. He suggested that the animal could be dead, which wasn't necessarily alarming, but the bear's location—in such close proximity to a busy hiking trail—raised eyebrows.

Carl and I drove up a slight incline along a winding dirt road that was restricted to the public. It took us through the forest, and I momentarily forgot the gravity of our mission. I cracked the window to inhale a whiff of the crisp mountain air, my mind lost in nature-induced daydreams. We traveled in silence, both of us enjoying how slowly we had to go on these roads, riddled with large rocks and interrupted by shallow creek beds, until we arrived at the entrance to the hiking trail. The air was cooler up there than where I'd parked, and I was pleased with myself for thinking ahead and bringing my fleece.

We found the trail and started walking, scanning the nearby landscape for the dead bear. I figured it couldn't be that hard to spot a large, dark-brown, furry lump on the forest floor, but I also knew this could be an all-day adventure.

Compared to many of the hiking trails in this part of the Sierra Nevada mountain range, this one was flat, and wide enough for Carl and me to trek along side by side. We hiked for a while, until the trees became denser and the forest began to feel a bit wilder, and I worried that we didn't have our bear spray or Karelian Bear Dogs for protection.

About thirty minutes into our hike, the trail meandered close to a small and almost-dry creek. I suggested to Carl that this could be the spring the caller had identified, and we stopped to take a good look around for the bear. Unable to see anything out of the ordinary from the vantage point of the trail, we decided to head off the path, walking along the creek bed. Our job quickly turned from field-based learning to wildlife forensics. All of a sudden, we were detectives, using clues and following our instincts to find a victim and determine the cause of death—in addition to assessing whether the entire bear population was under threat.

Carl spotted the bear first, and when my gaze followed his outstretched arm and landed on the dead animal, my heart sank. The death of wild animals doesn't bother me in theory, but something about the reality always causes me to need a moment. I took a deep breath, and we slowly walked toward the bear. The creek was maybe six inches deep, so the bear was hardly submerged, and in its stillness, the round hump of its back seemed especially large. It wasn't obviously dead, but there was no other explanation for it to be lying face down in the creek. Feigning fearlessness, we continued walking toward the bear, taking slow, careful steps. As we approached, Carl and I simultaneously began making loud noises as a test to ensure the bear was truly dead. Yelling our usual *"Hey, bear!"* and *"I see you, bear!"* in deep, resonant voices, the only response was the sudden quieting of the forest's chirping birds. When we were about three feet away, I stepped ahead of Carl and approached the animal, squatting down at its face.

I'd anticipated that getting up close to this dead bear would disturb me, but its face looked all too familiar: the same as the peaceful, sleeping face of the tranquilized animals I frequently handled in Tahoe. As I got closer, putting my hand in front of its partially submerged nose to see if I could feel any breath, I wished it were just taking a very wet nap.

Carl and I confirmed with each other that the bear was a juvenile, under three years old. Young bears, especially males, are more prone than mature bears or females to get into trouble. They typically disperse widely across landscapes while looking for new territory after leaving their mothers and, therefore, they have a greater likelihood of coming into contact with humans or human-modified environments that can be challenging and often deadly.

The bear had no pulse, so Carl and I pronounced it dead. We took a GPS point of the location and texted it to colleagues at NDOW, asking them to drive a special truck to our location, to pick up and responsibly dispose of the carcass. We snapped pictures of the animal the way it was found and recorded the approximate time we discovered it. Once the documentation was finished, we needed to drag the dead bear out of the water and answer our burning question: *What had killed this animal?*

We carefully stepped into the creek, measured the bear by width and length, and snipped a fur sample for potential DNA analysis. We then moved the bear, something I had done by myself and with Carl. Yet this time was notably different.

Carl grabbed the front paws, and I picked up the hind legs, shivering when I felt the cold, stiff limbs, so different from the warm and limber body of the tranquilized bears I typically handled. With great effort, Carl and I pulled the bear, which weighed about 250 pounds, onto the creek bank. We turned it onto its back, allowing us to identify its sex and exposing the wet fur on its round belly, and we examined it for any obvious injuries but didn't see anything.

Carl went into his backpack and pulled out two sets of bright-blue latex gloves for us to use and a small knife sharp enough to pierce the bear's flesh. He said we needed to comb through the bear's fur to look for bullet holes. This bear had probably been shot, possibly illegally, and we'd want to find the bullet, which might allow us to trace it back to the gun owner. With gloved hands, I took the top part of the bear and carefully searched for any signs of a bullet. After considerable effort from both Carl and me, we came up empty-handed.

Typically, when disease strikes the large-mammal community, the affected individuals experience a drawn-out illness, which often ends in mortality. And there are almost always physical signs, like an emaciated body, patchy fur, swollen organs, or irritated skin. Although we weren't ruling out the possibility of disease as a cause of death for this young bear, the case was becoming more mysterious by the minute. I don't have a background in forensics, and I don't watch many crime shows, yet I knew the next step: a necropsy. A necropsy is an autopsy for animals, a dissection of a corpse to hopefully determine the cause of death. I'd been under the impression that these things were carried out by trained medical professionals in white lab coats in cold surgical rooms late at night while detectives lingered nearby. But this was a black bear, and Carl and I were black bear experts, so it was up to us.

Carl already had his knife in his hand, but he reached into his bag and pulled out a smaller one for me to use. Again, we each took a section of the bear and inserted the knives, just enough to peel back the skin in roughly six-inch sections. As long as I'd been studying bears, from textbooks and scientific papers as well as in the field, I'd never cut into one. But my days of performing dissections in high school and college proved useful as I confidently detached the bear's skin from the rest of the tissue and tendons without piercing a single blood vessel. After a full forty minutes of skin removal and still no evidence of a bullet hole in the bear's body, we prepared ourselves for more serious "surgery."

Carl and I pressed on, now completely unsure of what we were looking for. We decided that before we looked at the heart or the brain—if that were even possible with the few tools we had—we should examine the gut organs, to see if something was obviously wrong with the bear's stomach. Not knowing where to begin, Carl handed me the bigger knife. With a deep breath, I made a slow incision, cutting through the now-skinless abdominal wall, and the pierced veins flooded the incision with blood. I sped up, using a sawing motion to cut all the way through the thick abdominal muscles, until I felt the knife pierce completely through. Carl and I each grabbed a side of the abdomen and pulled it apart, allowing an immense amount of blood to spill out, along with what seemed like far too many gut organs, much larger than what I'd expected. Thankfully, a crisp mountain breeze mitigated the offensive stench that emitted from the bear's gaping carcass.

Despite this being my first internal investigation, I confidently identified the inner organs: the liver looked like a liver—dark purple, flat, and triangular—and the gall bladder looked like a tiny bag filled with bile. I thought about how bears, especially Asiatic black bears, are often poached solely for their gall bladder, which can be used in traditional medicinal practices—the bile was especially valuable for treating disease.

The stomach came out next, and it was full. In fact, it seemed unusually enlarged, and Carl and I agreed we needed to open it to see what the bear had been eating. The entire necropsy had been messy and smelly, but opening the stomach and exploring the contents would take this process to another level. I braced myself, not wanting my own stomach to react to whatever we found. Carl made an insertion in the overly stretched stomach lining. He then held the organ while I inserted my fingers and slowly widened the hole.

Contents began to spill out, and nothing could have prepared us for what we saw: the young bear's stomach was full of ketchup packets.

Carl and I stopped everything. We removed our gloved hands from the bear's body and stared at the stomach, our minds at the intersection of utter disbelief and complete acceptance. Our personal understandings of black bear ecology and ecological theory, as well as what was known to science, had been made manifest.

This young bear had probably been exploring the edges of town, or perhaps dispersing through the area en route to finding new territory, and had become drawn to the smells of a fast-food restaurant. This bear might have made a habit of rummaging through the restaurant's garbage cans and gotten a taste of the delicious, high-calorie food scraps. It also was possible that this was the animal's first time scavenging in this area, and his fatal mistake would be his last.

Ketchup is sweet, and it's every bear's instinct to ingest as much sweet food as they can get their paws on. A sweet taste indicates a higher-than-normal caloric load, which is critical for bears as they enter hyperphagia in the fall and prepare their bodies for hibernation. This is one of the most well-known elements of black bear ecology. Whether this young male bear was a habitual dumpster-raider or he died on the first try, the fact of the matter was that human negligence had killed him. A basic bear-proof garbage can would have prevented this untimely, and likely painful, death. The crucial nature of our study about the ways human influence on the environment impacts black bears was presenting itself in a way we could not ignore: humans were rapidly decreasing the quality of the black bear habitat, and bears were the victims.

Carl and I had solved the mystery, and although we'd cycled through the emotions of this bear's death, something about the ketchup packets, and knowing the extreme suffering the young animal must have endured, caused new feelings to surface. As we used our smartphones to take pictures of the stomach contents, I began to feel that my work as an ecologist wasn't doing enough to solve the problems between humans and bears. Sure, conducting a long-term study moves the scientific field further in important ways, but many people in Tahoe were exasperated with bears, and bears were risking a lot by living in a habitat with high human populations. I felt called to learn more and to communicate what I knew about human-bear coexistence with as many people as I could. I wanted to make sure that his death, and the death of all bears at the hands of human neglect or ignorance, wasn't in vain.

During my times in western Nevada, some of my colleagues became my dear friends. Not only did I learn a great deal about black bear biology, management, and policy from Carl, but I spent considerable time with his family, friends, and coworkers.

I was able to indirectly absorb important information about the female empowerment embedded in the local culture from another colleague, Mark, and his wife, Jenny. Although I pride myself on being open-minded and well-traveled enough to know I shouldn't enter a new community with preconceived expectations of the cultural norms, in the rural mountain towns of western Nevada, I was surprised by how strong women are perceived to be. In this part of the world, women are hunters, and nobody bats an eye when a woman is the main hunter for the family.

Enjoying a meal at my colleague's home was a regular occurrence, and the generous invitations to family meals became my favorite part of the local culture, especially coming from the vast cities of the East Coast, where ordering takeout was my main style of dining.

Sitting down at the family table one night, my taste buds awaited what I was sure would be another delicious dinner. Although both Mark and Jenny are skilled subsistence hunters and cooks, the family dinners that I happily attended consisted of Mark cooking whatever meat Jenny had hunted. This routine occurred without any discussion or acknowledgment that this arrangement was anything but normal, and I found it fascinating and progressive. After my first meal with the family, which also included their children, I realized the wife hunting and the husband cooking wasn't the norm solely for them—it was commonplace across much of the community. With that said, Jenny was particularly badass, which is another reason I admired her so much.

Although I'd eaten many different kinds of game meat during my years of fieldwork in Eastern and Central Africa, outside of the venison jerky I distinctly remember a classmate bringing in during an elementary school show-and-tell, I hadn't had game meat from the US until my Nevada work. In the rare cases that we weren't eating a recently hunted ungulate, like elk or venison, the family would default to their beef stash in the garage freezer. I never knew what to expect, other than a fresh cut of meat with a good story as a side dish.

One evening after an uneventful day of routine meetings with the wildlife department, I returned to Mark's home to spend some quality time with the boys and the dogs and was promptly invited to stay for dinner.

Jenny had done the hunting and had taken an extra-special trip for this particular objective. I continued playing with the kids and the dogs as the aroma of sautéed mushrooms and butter began wafting from the kitchen, making my mouth water. When the time came to sit down at the dinner table, I busied myself with placing salad on my plate and convincing the kids that they, too, should eat their vegetables. I passed around the bread basket and the vegetables that had come off the grill and, at last, I was able to place the meat onto my plate. As I devoured thin cuts of meat smothered in a golden-brown mushroom gravy, I noticed Mark and Jenny staring at me, waiting for a reaction.

"This is amazing," I said, and they grinned at each other.

"Do you know what we're eating?" Jenny asked. I realized that I hadn't asked her what she'd returned with or what made this trip so special. Had she hunted with a bow and arrow instead of a rifle? Was the trip to a special part of the state that she rarely visited? Had Mark learned a new recipe that made a relatively common type of game meat especially tasty?

"It's mountain lion," Jenny revealed.

I stopped mid-chew, almost gagging on the meat as she said the words.

"I'm sorry?" I managed to ask while taking a slow and deliberate swallow.

Mark and Jenny continued to stare at me, looking for the astonishment and approval that they'd expected at what, to them, was an impressive feat. Here we were around the same dinner table, filled with love and caring and friendship. Here we were, people from the same country who speak the same language and had spent lots of professional and personal time together. Here we were, so similar yet so different.

As a nonhunter and former vegetarian, I'd gone through a tremendous personal journey toward decreasing, and eventually eliminating, my judgment around people who hunt wild animals for food. Although I don't pressure others to join my line of thinking, I personally believe that hunting one's own food is a reasonable thing to do. Yet something within my soul quivered at the idea of hunting and eating a large carnivore, like the mountain lion meat sitting on my plate. Mere seconds before, it had tasted delicious and fresh, and now it made me want to cry.

My friend who had hunted this animal took immense pride in the achievement, because killing a mountain lion was indeed a notable feat. They are solitary and elusive animals, difficult to find whether you are looking for them or not. Even amid my black bear study in Nevada, we'd found evidence of black bear–mountain lion interactions, but the only way we ever saw mountain lions was from the images on camera traps. As quickly as they would arrive at the scene, they would be gone again.

"Wow, it's delicious!" I said, and it was the truth. The mountain lion did taste great. I'd never eaten cat before, but the meat's light flavor and pleasant texture wouldn't have led me to guess that's what we were consuming.

The conversation pivoted to the tales of Jenny's hunting escapades. Although mountain lion hunting tags are available to most Nevada residents, many people choose not to attempt the hunt because of the time and effort involved. Her story from this hunt was one of adventure and a close knowledge of the outdoors, leading her to shoot a large male cat that would feed the family for a long time.

Perhaps one of the most interesting things to emerge from our conversation over the mountain lion dinner was the fact that, in spite of being avid hunters, Mark and Jenny had no interest in hunting black bears. They'd heard that bear meat was difficult to cure, and the recipes were not especially good. Neither of them would divulge this, but I could tell they drew the line at hunting black bears because of the intimate relationship many of us—community members and researchers alike—had built with these animals over time. Handling black bears every day, working tirelessly to protect them and keep their populations thriving, stuffing newborn cubs into our jackets during winter den surveys—for my colleagues and me, black bears were special.

TWELVE YEARS LATER, I've captured, processed, and followed countless black bears through the Sierra Nevada mountain range, Rocky Mountain West, Canada, Pacific Northwest, and now the California Central Coast. I have sat in volatile community meetings both in defense of bears and in defense of the people who live among them. I have discussed bear ecology and bear conflicts on television, in newspapers and magazines, in public lectures and academic seminars, and now in book form. This has all been possible because I have been mentored by some of the top black bear experts in the world, a privilege I remain eternally grateful for.

Although I may be guilty of having an extreme fascination with black bears, it's difficult to find a person who isn't excited about bears in some way. Bears have cultural importance in many communities around the world. In America, children from all parts of the country fall asleep to stories about bears and their fictional adventures. Bears are arguably the only carnivores portrayed as nonthreatening, endearing friends in children's stories. My own childhood bear, aptly named Teddy, traveled with me when I left home for college and has been a part of each move since.

Bears also serve as symbols, showing up as mascots for popular sports teams as well as the main characters in children's television programming. In many parts of rural America, bearskin rugs retain the same prideful appeal today as they did hundreds of years ago. Historically, bear parts were respectfully used as adornment in some indigenous cultures, and even today, many hunters I have come to know cook bear meat from recipes passed down from generation to generation.

And it is not only bears in the abstract that are so dear to us as North Americans. Viewing bears in the wild is one of our continent's most popular forms of ecotourism—an industry that's an important revenue source for states that trickles down to taxpayers' wallets. Black bears are one of North America's last remaining large carnivores that are still abundantly viewable in their natural habitats, and nature lovers travel from far and wide for the chance to watch a bear catch salmon in a running stream or scratch its back against a sturdy tree trunk.

In spite of people's endless fascination with bears and other creatures, human beings are sensitive about having to change their own behavior or routines in order to accommodate wild animals.

Some conservationists feel that there's an inherent value to wild animals—that because they exist on this earth just like we do, they have the right to survive and be taken care of. Other people might cite other reasons. I often find myself using the argument that bears are important for ecosystems. If we want healthy, functioning ecosystems that give us clean air, water, and resources, we need bears for that. In places where we have an absence of bears—where they used to be but are now gone—there's something missing, an element that technology can't always make up for. There is something to be said about having intact, thriving wildlife communities, even when this is inconvenient for humans. These animals help us save money, save resources, save time, and dwell in healthier spaces.

We all live and cohabitate together, and parasitic relationships are doomed to fail. My discoveries in Tahoe began alerting me to the negligence of those outside my field of study and catalyzed my work in environmental activism. I was forced to question my privileges and put to good use my education, to protect those who couldn't otherwise protect themselves.

I cared for animals and for humans, and I longed to marry the two seemingly opposing passions into one streamlined career. With brick and mortar laid, I vowed to begin the work of bridging my two dissonant paths.

Chapter 7

T he bush was as beautiful as I remembered it from my first visit nine years prior. As our bus drove from Nairobi into the wilderness, the sights and the smells flooded back to me. I smiled when, outside the driver's seat window, I spotted a marabou stork—the same animal that had welcomed me to Kenya on my first international experience.

This time, it was 2014, and I was a PhD candidate in the fourth year of my program at Columbia, returning to Kenya as the faculty lead on an undergraduate study-abroad trip. No longer was I the naive city slicker who'd never been on a hike. I was now the resident expert on all things outdoors.

Shortly after I glimpsed the stork, the back of the bus erupted in excited clamor as the students noticed it for themselves, their hands pressed to the glass of the bus windows. For many of them, this was their first glimpse of East African wildlife.

My journey had come full circle. I was back in the place where my love for wildlife ecology had blossomed, with students who were bound to experience the same epiphany. I'd still be living out of a tent—the accommodations don't improve with seniority—but I was responsible for more than myself.

Far different from my previous experiences in East Africa, this trip centered not on the study of one particular animal but on surveying the landscape and its wildlife. The students and I would participate in game drives and spend afternoons studying, observing, and noting the patterns of the land and its inhabitants. For the first time, I had the opportunity to take in the grandeur of Kenya without the worries or high stakes of data collection. For lack of a better term, I could be a tourist.

Even so, the lingering shadow of a recent, unexpected blow darkened my trip. My paternal grandfather, George Grant Jr., passed away a few days before I planned to leave for Kenya.

I'd been in constant contact with my father's side of the family. My grandfather knew I had a chance to teach in a study-abroad program in Kenya, and though he shared in my other relatives' pride, his health was rapidly deteriorating. So the question became whether he'd pass before or after I boarded the plane for Kenya.

Few things can make you give up on your dreams the way grief can. After my grandfather's passing, things that once seemed so important to me—like the prospect of experiencing my career come full circle—meant little. Unfortunately, it was too late to cancel the trip. After a heavy and emotional conversation with my family, we concluded that I shouldn't halt my life because of my grandfather's death.

So there I was, once again admiring Kenya's flora and fauna, though my heart felt heavy and everything seemed less vibrant. I kept my loss to myself. Even though I'd left with my family's blessing—and my grandfather would've encouraged me to go as well—I still felt ashamed and guilty about having gone forward with my plans. This wasn't some once-in-a-lifetime opportunity that would provide immense financial or professional benefit. It was just something that excited me, which I knew would bring me joy. But that didn't feel like an adequate justification to miss his funeral.

During those first weeks, it was a thrill to witness the students experiencing the bush, but my grief hung around me like a thick woolen cloak. At the end of one particularly draining day, I was sitting with the local Maasai chief, a man I had come to know over my years of work in the region. During my first trip to Kenya, I'd benefited so much from my interactions with the Maasai that I wanted to ensure the students I was working with could have the same kind of contact with the community. Several leaders had an open door to our field site, and they'd come daily to teach about their culture. This particular chief spoke English beautifully and became a short-term friend. After each class he taught, we'd sip tea together and chat.

That day, he noticed that I seemed drained and politely asked me where my mind was. When I opened my mouth to speak, no words escaped. Tears streamed down my cheeks, and I flushed with embarrassment. It was the first time I'd allowed myself to cry since arriving in Kenya—the first time I'd allowed myself to think about my grandfather. Mistakenly, I'd believed that withholding my tears would suppress the tidal waves of grief buffeting my heart, mind, body, and soul every day.

In a moment of vulnerability, I told the chief I was mourning the loss of my grandfather. I told him that I felt guilty—guilty for having chosen a personal and professional opportunity over the chance to honor my grandfather's life and legacy.

"Why can't you honor him?" the chief asked.

I explained the way we usually mourn in America, how the family of the departed comes together to honor their legacy. We tell stories and eat food and go to a church and pray, and then we bury the body in the ground. I told him that in America, we honor the dead by having a funeral. And I had missed my grandfather's. I had not had time to mourn.

The chief looked at me.

"We do that here too," he said.

I was being selfish, he agreed, but not for any of the reasons I'd mentioned. I was not selfish for having chosen the field course over the funeral but for having not figured out a way to honor my grandfather independently while in Kenya.

"Let us help you," he said. "We will bury him here."

Under normal circumstances, I might have declined, but by then I was at such a low point I was open to anything that might alleviate my pain.

I'd been privileged to have my grandfather in my life well into adulthood. George Grant Jr. was born in central Texas in the early 1920s. I knew him as a gentle giant. At over six feet tall, he was slim and lanky, yet coordinated. His Texas roots shone through his slightly Western style.

Although he was Black, he looked distinct from every other member of my family. Rumor had it that he was biracial, but his paternity was a bit of a mystery. My grandfather had light skin with a reddish tint, straight hair, and light-brown eyes that were soft and gray when I knew him in his older age.

I called him Hop. As the story goes, when I was learning to talk, it was hard for me to pronounce "Grandpop," his preferred title. The closest I came was "Hop," and the name stuck.

He died in the same house he'd helped build in the 1950s, when my grandparents moved from central Texas to Redwood City, California. They were part of the Great Migration of Black folks who escaped the violent Jim Crow South to find work and safety in "progressive" states. He'd taken construction jobs, and he and my grandmother became founding members of the Jerusalem Baptist Church in Redwood City. When Asa and I were children, we spent many Sundays at Jerusalem, listening to my grandfather preach and my grandmother play the organ.

He hadn't gone to college, and racial trauma had dominated much of his early life. Despite his background of struggle and sacrifice, out of all four of my grandparents, Hop was by far the most supportive of my nontraditional interests. "How are the bears?" he'd say to me as soon as I'd enter his house, followed by a big hug and kiss. His final few years were marked by diabetes-induced health issues. He slowly began losing his limbs, a common occurrence with long-term diabetes, which changed him drastically. Still, his Christian faith and love of Scrabble kept him in great spirits until the end. I missed him when we weren't together, and I missed him even more after he died.

THE MORNING AFTER my conversation with the Maasai chief, I awoke before sunrise and strode to the edge of the village. There, I met the chief, his wives, and two elders, who adorned me in traditional Maasai red cloth and wiped red paint across my cheeks and forehead. I walked in silence between them down the road as they sang and chanted in Maa until we arrived at a baobab tree that was nearly one thousand years old. One of the elders got down on his hands and knees. He dug a small hole at the base of the trunk. When he finished, he stood again and instructed me to kneel where he had once been.

As soon as my knees hit the ground, I began to cry. My vision blurred as memories of my grandfather flooded my mind. Behind me, the chief tilted my head to the ground so my tears dropped into the freshly dug hole, wetting the soil.

"You exist because your grandfather existed," said the chief. "Your tears are a part of him. And today we bury them."

Slowly, my crying ceased as the mourners helped me to my feet. The elder filled the hole and patted the ground. As I stood, I felt taller. Lighter. And forgiven.

All of the mourning party turned away from the baobab tree and returned to the village in silence. We arrived at the camp as the early morning sun's first rays kissed the bush. I made eye contact with the elders and said, *"Asante,"* a sincere and heartfelt thank-you. They departed, and I entered my tent to prepare myself for the day's classes. I'd told the program staff I'd been invited to spend the morning with some of the Maasai villagers, but I didn't explain why. As often seemed to be the case in my fieldwork, I was isolated in this experience—though this time, it was by choice.

Before I left Kenya, a baby boy was born in the village. I would learn that they named him George.

This was the highest honor, and one that my grandfather would have cherished. I lamented that he'd never know this baby was carrying on his legacy, but that was the point: he didn't have to be alive to be continually honored. Baby George was a reminder that life carries on and remains hopeful and positive and joyful. And I felt privileged to be a part of a Maasai community that consistently reminded me of that.

ONCE THE STUDY-ABROAD program ended, I was free to do some traveling and exploring. I planned to stay on the continent as long as possible. I knew New York, my graduate program, and my more traditional life as a wife were waiting for me. But I loved Africa—the way it made me feel, the way it smelled, the red dusty soil, the people. Being there reminded me that I was familiar with a part of the world that most Americans never would be. I had lived in Kenya in 2005 and in Tanzania in 2009, and there I was, five years later, returning to the continent. I wanted the identity of a wildlife ecologist with an expertise in East African wildlife. Tahoe was in my heart and soul, but East Africa felt like home in many ways.

That trip, I visited Tanzania, Zanzibar, Rwanda, and Burundi. The latter, a small East African country not well-known to me at the time, is unique. Burundi experienced the same genocide as Rwanda in the 1990s (the two countries share a border), but it wasn't internationally recognized. And while Rwanda has since experienced a lot of economic stimulus, Burundi has been left behind in some ways, remaining underdeveloped compared to many of its neighboring countries. But that certainly takes nothing away from its beauty and potential. The country is one lush, grandiose, vibrant green ecosystem.

I'd gone to Burundi to help my best friend from Emory and DC, Nora. After she left her job at WWF, she'd moved to Burundi with her husband to support the launch of a medical clinic and lead the early childhood education efforts through an organization called Village Health Works, a partner of UNICEF. VHW was founded by a Burundian man who had traveled to the United States to solicit funds for this huge undertaking and had raised enough money to begin building the clinic in a remote part of his homeland.

Nora and her team had found the perfect place to build a preschool alongside the clinic. She'd always been interested in international development, with a focus on establishing preschools in developing nations. At the time, Burundi's children would most likely end up working on their parents' farms, which was dignified work but typically done out of necessity, not choice. The philosophy was that if we provided these children an education, they'd have a higher likelihood of evolving into powerful, positive leaders and help shape strong futures for themselves, their families, and their country. Educating the next generation would give the country potential to grow exponentially.

Nora's program began formal classes for the children in the newly built classroom, adjacent to the medical clinic that was being built in the village. I visited briefly to add an environmental science and climate change element to the education that some of the teenage community members were receiving. My equipment included a projector, so I'd use PowerPoint presentations as part of my lectures. After the lessons, the students were always eager to ask me questions about American pop culture. On several occasions, students would raise their hands and ask me if I knew Beyoncé.

For myriad reasons—including language, geography, fiscal barriers, and discrimination—Burundians were isolated from much of the world, especially in terms of opportunities for education. Most people I encountered had minimal formal knowledge of ecology or climate change, despite living in such a rich, untouched ecosystem. With the help of translators, during casual conversations out in the field or walking to and from villages, we'd ask what changes they'd noticed in their environment and gauge their general awareness of climate change. Mostly, we discussed the types of agriculture they were skilled in and that were part of their cultural background, and we found crops that were both culturally valuable and climate-resistant—for example, ones that could withstand periods of drought or flood. We also talked about ways to reinforce their homes, so heavy rains or fires would be less of a threat.

Whereas many agrarian cultures can draw knowledge from a deep well of generational wisdom and traditional practices, Burundi's population was young. The genocide had left a generation of children who grew up largely without parents or grandparents. Most of the community members I interacted with were my age or younger, and many were parents to young children. Nevertheless, they were successfully rebuilding their country after tragedy and largely without the guidance of elders. In the context of those traumas and challenges, managing the present was a much more pressing need, rather than focusing on how climate change might affect the future.

TEACHING ABOUT CLIMATE CHANGE made me realize how pervasive its effects are. These specific people had barely released a fossil fuel in their lives. They had no hand in contributing to global climate change. And yet they would suffer the worst from it and have to enact the most drastic changes to survive its impact. The world has failed them in this way, and it was my job to explain the injustice of it all while also helping implement practical solutions—Band-Aids for bullet wounds.

Today, I can better articulate things I couldn't back then, when I was just beginning to find answers to my own moral questions. That year, it became abundantly clear to me that humans are far more important than animals: if humanitarian issues are at hand, they must be solved before unpacking the ecological plights of our time.

The issues plaguing wildlife intersect with many contemporary issues, such as humanitarian aid and economic justice for marginalized peoples. Conservation efforts don't have a future without an economic component. Poverty is like a rain cloud—it dampens everything underneath it, including how we deal with and interact with our ecosystems. Investing in conservation takes money, and many communities in the most at-risk areas don't have the financial means to implement sustainable solutions.

Nearly a decade prior, during my first visit to Africa when I told the Maasai people about the legacy of American slavery and the Black experience, I had been overly optimistic. My grandfather lived through the worst of Jim Crow and the best of civil rights. As a Black American in the twenty-first century, it's easy to buy into the promise of social progress. When we look at developing nations, areas that I've visited numerous times throughout my work in the field, it's easy to believe that the inequality seen in such communities would never and could never exist in the United States. But the reality is that disparity exists in our own backyards.

As a community, we must do better for one another. Only through collective work can we heal the wounds and traumas of the past and strive toward a better future. The simple act of the humble Maasai mourning and grieving my grandfather's life reminded me that while we hurt together, we also *heal* together.

I brought this same philosophy to my brief work in Burundi. My grandfather and his generation left a better world for me and my Black peers, but by no means a perfect one. Through my work with animals, ecosystems, *and* humans, I want to leave the world a little better than I found it.

Chapter 8

After working in the remote evergreen forests of Lake Tahoe and the depths of remote Burundian villages, I returned to New York City at the end of summer 2014 to write, finalize, and eventually deliver and defend my dissertation.

Once upon a time, the land and water the city now occupies were stewarded by the Munsee Lenape people. Lenapehoking, the traditional name for the land, as given by its original inhabitants, was formerly home to a vast array of birds, amphibians, reptiles, fish, and mammals—including the black bear.

About two hundred years ago, the range black bears occupied was widespread. But in most US states, these bears were over-hunted and sometimes entirely extirpated. Since then, people like me have been working hard to get these populations to rebound, with a good amount of success. For my dissertation, I wanted to pinpoint the current threats to black bear populations, starting with my case study in the Lake Tahoe Basin. What was threatening them? What human activities were dwindling these populations to a point of inevitable demise? And what did the safe-haven habitats have in common with each other?

Through my years of fieldwork, I observed some undeniable trends: the majority of bear-mortality incidents were occurring in forested regions, and the second-most-common incidents were happening in dry scrub brush, areas not very populated by humans. This highly contradicted everything we'd previously thought.

We'd believed that bears were more likely to be killed in places with large human populations. While the data I collected showed that bears were being killed in rural communities, it still correlated with our hypothesis that humans were impacting bear mortality—that is, they were being killed in vehicle collisions or put down because of public-safety concerns. Curiously, there was also a discrepancy between the sexes of the bears. Male and female bears were getting hit by cars fairly equally. But wildlife authorities were being called in more often for male bears to be put down for public-safety reasons. Male bears were also more prone to being hunted for sport. I began developing statistical models to analyze the data and determine the disconnect. My model was able to predict, based on geographical data, where bears would be more likely to die across the landscape. I made maps of hot spots of black bear mortality to help wildlife managers on the ground handle the populations.

Impressed by my work and convinced that presenting and networking at conferences would prepare me for my defense, my dissertation committee suggested that I apply to the world's premier ecology conference. I did, and I was invited to present my working thesis. After struggling in the department, I felt that I'd finally been legitimized. It was a momentous occasion.

I landed in San Francisco and headed to my grandparents' house, where my field vehicle had been parked since I'd left Tahoe earlier that summer. I gave them both a quick hello-and-goodbye kiss then hit the road to Sacramento, where the conference was taking place. I turned on NPR for the drive and was inundated with reports of an eighteen-year-old Black man being fatally shot by a white police officer in Ferguson, Missouri. The young man's name was Michael Brown. Ferguson had erupted in protests, and the nation was again faced with civil unrest in demand of justice.

I was shaken by the news, but I was presenting first thing the next morning. I turned off the radio and tried to shut out my thoughts.

When I arrived at the hotel, I sat staring at the empty PowerPoint slides on my computer screen, unable to do anything but cry, refresh CNN, and cry a bit more.

One of my initial exposures to police brutality was on TV in the early 1990s, when I watched the footage of Rodney King being beaten by Los Angeles Police Department officers. My family and I had some hard conversations about that event, which sparked riots after the officers were eventually acquitted. While the situation in Ferguson continued unfolding, I couldn't help but remember Trayvon Martin, who had been killed the year before, and the fact that his murderer remained free. And of course the encounters and harassment my father and Asa had endured from the police also simmered to the surface.

At Columbia, I'd joined a group called SOCA (Students of Color Alliance), which served as a space to help process racially traumatic events. I did advocacy work with the group, demanding protection for Black folks within Columbia and in the community. As passionate as I was about science, I was a human being first—one who cared about social justice more than anything.

The next morning, when I got onstage, I gazed into an audience filled with friends, professors, and some of the wildlife ecologists whom I admired most in the world.

But my mind was elsewhere. I stumbled over my words and lost my train of thought. My energy was flat. My enthusiasm was sparse. And my will to present on something once so important, now so seemingly inconsequential, was apparent.

Did the people in the audience care more about the bear skull behind me, with a bullet hole through its parietal bone, than they did about the six bullets that had killed Michael Brown the day before? Did they care more about the lives and deaths of black bears than they did those of Black boys and men in the United States? The audience wanted me to talk about the bear's death, and it wasn't lost on me that I was giving a presentation about black bear mortality risk in an era when Black male mortality risk was much more important. Or at least it should have been.

I bombed. Big time. And I felt that I had shamed my department and embarrassed myself.

After my presentation, I walked off the stage, left the conference, packed my bags, and drove all the way back to my grandparents' Bay Area house, half lying to them about my time at the conference ending sooner than I'd expected. I changed my plane ticket to leave the next morning.

It was my own form of protest, ditching that unwelcoming and toxically navel-gazing space. I was ready to go home to Harlem.

"BABE, CAN YOU TALK for a sec?" I asked Oba. I was sitting on the toilet in the bathroom of the tiny Harlem apartment I lived in with my husband, the phone pressed to my ear.

"Yeah, what's up?" Oba was working at his law office down in the Financial District.

"So I randomly took a pregnancy test, and it's positive," I said, emotionless. "Isn't that weird?"

"Are you serious? Oh my gosh!" The only other time I'd heard him this excited was when he was watching sporting events.

"Wait, wait—don't get excited. I'm not pregnant. I'm just telling you because it's so crazy that these tests can give a false positive." I couldn't possibly be pregnant. I didn't *feel* pregnant. If I were pregnant, wouldn't I have known it?

He was confused but deferred to my judgment for the time being. We made a deal that when he came home from work, we'd do another pregnancy test together. I was certain it would be negative. I was in complete denial.

"Listen, babe, don't get your hopes up or anything because I'm definitely not pregnant. So it's going to say 'negative' when we do it tonight."

It was August 2014, and I'd deliberately gotten off of birth control pills several months earlier. While we'd tried to conceive, I stockpiled pregnancy tests in our bathroom so I could know right away if we were successful. I downloaded a period app so I could track my cycles. I became hyperaware of my body, and I thought that every tiny twinge was an early indicator of pregnancy. If I were somewhat tired or ravenously hungry, I'd rush to the bathroom to pee on the stick, only to see a negative result. After a few months of this, we stopped trying as intensely and intentionally. I genuinely felt that it would simply happen later, and I transitioned from obsessing about getting pregnant to obsessing about finishing my degree program.

That morning, I'd looked at the tracker because I figured my period would be showing up soon. The app indicated that I should've had my period several days prior. I hadn't felt pregnant and didn't have a clue that I was. It's often said that it's easier to get pregnant when you're not actively thinking about it, and that certainly proved true in my case. I'd been so immersed in writing my dissertation I hadn't given pregnancy a second thought.

I was twenty-eight years old, Oba was thirty-nine, and although I'd never been enthusiastic about starting our pregnancy journey, I agreed that I wanted to have children. I'd done some self-exploration and determined that although I wasn't super interested in babies, I could happily envision having school-age children whom I adored.

I wasn't opposed to getting pregnant in my late twenties; I wasn't opposed to doing it in my early forties. Ironically, as someone who was almost always planning everything down to the minute, I was open to however my motherhood journey played out.

And honestly, I figured that it would take a long time to get pregnant—possibly years. I figured this was the case because of the combination of being on hormonal birth control for over a decade, having a stressful life as a doctoral candidate, spending tons of time in the field, and knowing the pièce de résistance: our sex life was inconsistent, between my travel, his late hours at work, and my hyperfocus on dissertation writing. But we knew that many couples went through similar droughts during stressful life seasons, so we anticipated things would eventually pick back up.

That evening, two lines showed up on the pregnancy test.

"Holy shit!" Oba exploded with glee.

My denial was dislodged by shock waves reverberating throughout my body.

"Wait, are you even happy about this? Rae? Isn't this what we wanted?" Oba, who usually wasn't so emotive, looked genuinely concerned.

"I just . . . I didn't think it would happen so fast. I just . . . ," I stammered, trying to find a reason for my paralysis. "I thought I'd feel more ready." And then, as the shock subsided, the guilt began taking over.

Was I a bad person for not feeling ecstatic? I wasn't sad that I was pregnant, but I was surprised and scared. Part of me was used to having some kind of control over what my body was doing. But with pregnancy, so much is out of one's control.

"But I was drinking last night!" I started crying. "I had wine and tequila and I barely ate anything!" At the sight of my abundant tears, Oba's initial worry evolved into mild panic. He stared at me, still sitting on the toilet seat where I'd taken the pregnancy test, sobbing over the fact that while we had planned a pregnancy—and had the privilege of achieving one—I was freaking out.

The next day, I called my ob-gyn's office. Thinking the receptionist would insist that I be seen straightaway, I couldn't believe that her response was a stoic question: "When was the first day of your last period?" I told her the date, and she indicated that I was only a few weeks pregnant. The doctor didn't need to see me until I reached the eight-week mark, which was more than a month away. I scheduled the appointment, hung up the phone, and googled "what to do when you're pregnant."

Was my instant need for more information the scientist in me, or was it the fact that I'd been a perpetual student for almost my entire life? Was it because I was the first one of my closest friends to get pregnant? That there hadn't been a baby born in my immediate family since my brother in 1987? I can't explain the marching orders I sought, but I needed guidance, support, instructions. Google told me to lead a generally healthy lifestyle; eliminate drugs, alcohol, and tobacco; and see a doctor if I had any questions.

In spite of my efforts to have a healthy pregnancy, I ended up with hyperemesis gravidarum, which is basically an extreme form of morning sickness that happens all day, every day. The cause of this ailment is unknown. I spent my pregnancy constantly vomiting and lost about twenty pounds, a dangerous situation for both me and the baby. My symptoms were so severe that I even landed in the hospital a couple of times and, at one point, had to take a leave of absence from my studies, even though I was close to finishing my PhD. Since I spent most of my days throwing up, I was extremely isolated, and to add insult to injury I felt disgusting. I still struggle to think back on it. Contrary to what I thought the magical time of pregnancy would be, my experience was nothing less than traumatic.

THE SOCIAL JUSTICE call to action that I'd experienced when Michael Brown was murdered burst forth anew when I was late into my pregnancy. As I sat in the fifth-floor apartment Oba and I shared, I heard chants from the window. Despite the lateness of the hour, Oba was still at work, and I decided to venture out to see what was going on.

I had an inkling about what was happening, so I opened Twitter and searched #BlackLivesMatter to find information about current protests. Since Michael Brown's murder two months prior, several more had followed, most recently the murder of Tamir Rice in Cleveland, Ohio. He was only twelve years old when the police killed him, claiming they mistook his toy gun for a real weapon.

I carefully descended the five flights of stairs and headed into the cold streets of Harlem, clutching my phone in my hand. We lived on 126th Street, so it took only moments for me to walk to Harlem's famed 125th Street—often described as the Black Broadway—a throughway in Upper Manhattan that is rich with Black American history, business, culture, and success.

Although it was evening, the streetlights blazed and illuminated the hundreds of Black Lives Matter protestors, holding signs and chanting, "Whose streets? Our streets!"

I stood on the corner and watched, joined the chants, and shed tears. At one point, I ventured into the protest and started walking with the group, but I was scared. I worried that the protest would be met with aggression from law enforcement—tear gas, rubber bullets, arrests—all the things I'd seen on the news in the years since Trayvon Martin was killed in 2012. I acknowledged that it wasn't just me who could potentially get hurt—my unborn child would also be endangered. All I wanted was to join this community of brave people putting their safety on the line to demand justice and accountability for the racial violence that continued to plague our country. But I didn't want to risk getting hurt, and I had to think about more than myself.

Two blocks into the march, I made my way to the sidewalk and slowly began walking home, to continue monitoring the protest from the perceived safety of my apartment.

Several months later, as I defended my dissertation, I was *pregnant* pregnant. I, a young Black mother, wobbled into a room of older white advisors, the stark dichotomy striking me. Remembering the failed presentation at the ecology conference in Sacramento, I approached the defense with a calculated, unemotional dedication to fact, to the pragmatism of science. As I spoke, I rubbed my tummy out of habit—and maybe as a way to calm my nerves.

I gave my dissertation defense, an hours-long presentation that covered everything from the unique ecology of black bears in the Lake Tahoe Basin to the statistical modeling techniques I used to calculate which landscape characteristics influenced their ability to thrive or their risk of death.

Although the road to the defense had been rocky, particularly because my pregnancy journey had been traumatic at times, I was well-prepared. I'd paced myself, asked for help, practiced, revised, and finally presented. The questions from my dissertation committee were challenging, but I remained calm and worked through them successfully. I'm sure that part of my serenity came from having my baby girl almost fully grown inside of me. As a biologist, I knew that any stress hormones I felt would be received by her as well, so I chose to be steady. The end result was a win for both of us.

I answered my final question and was excused to the hallway by my committee, so they could discuss what my fate would be. After deliberating for fifteen minutes, they welcomed me back into the room with a round of applause and the terrific news that I'd passed.

I was ecstatic, but not necessarily relieved. The whole experience—all five years of the PhD program—had been hard work. I was burned out, and that feeling rose to the surface, eclipsing my success. I needed rest, and possibly a reset. And more than anything, I needed to prepare to give birth.

I wobbled out of the room, and two weeks later, I gave birth to my firstborn, a beautiful baby girl whom we named Zuri. Her name, the root of the Swahili word *mzuri*, means "good," and it was a nod to the love I held for East Africa.

Around that time, I read a peer-reviewed journal article that attempted to explain why Black men are killed so often. In the United States, unarmed Black individuals are about 3.49 times more likely to be shot by police than unarmed white individuals. This is unrelated to crime rates, even when accounting for race-specific crime rates. The science laid out a truth I had been dreading: Black people are killed more often, for no other reason than racism.

The article reminded me of an academic piece I'd read a year prior that attempted to explain why unarmed Black men were killed so frequently and unjustly. The piece was transformational for me. This article wasn't in a social science journal or a Black studies journal, where one would expect to find such topics covered. Rather, I found it while searching for literature to support the statistical methods I was using to analyze my carnivore-movement data for my PhD dissertation. The article sat at the intersection of ecology and statistical methods applied to sociodemographic data. It was written by a white author who wanted to show fellow ecologists that police violence aimed at the Black community is an issue that can be viewed through the same lens that a traditional academic ecologist might use. I interpreted this approach as saying, "If you can understand these statistics when it comes to environmental risks for wild animals, then you can understand these statistics when it comes to social risks for Black Americans."

I read the entire piece rapidly, rushing to reach the conclusion, hoping the author would reveal whether social variables explained why unarmed Black men were murdered so often by police. I worried there would be bad news. Maybe the author found some unfortunate trend or pattern that suggested guilt on behalf of the Black men involved. Yet the paper stated that the only pattern found was one of racial bias in police officers—a bias that makes them more inclined to pull the trigger when the person in question is Black.

In short, anti-Black racism is deadly. Just as Black folks have been voicing for centuries.

While I feel slightly ashamed to admit it, the author's whiteness was validating for me. It showed me that conversations about Blackness and racism belonged in professional scientific spaces—that exploring the intersectionality of science and social justice was crucial, vital work.

I was finally seeing a non-Black scientist who cared. And cared a lot. A non-Black scientist was suggesting there was an intersection between social justice and science and that it was time to take the matter seriously.

It was time for me to elevate my thoughts and feelings to the next level of action. It was time for me to say out loud that I was unfulfilled using my effort and energy figuring out how to reduce incidents of black bear deaths when the lives of Black people were under threat. I needed balance between my ecology and social justice work, and that needed to start with more acknowledgment, action, and allyship from my scientific community.

As difficult as it was to swallow the truth that anti-Black racism and police violence plagued the Black experience in America, I began to find some semblance of solace in the science. I was a new mother, and though it was hard to stay at home with my young daughter, away from the protest-filled streets, I knew it was the responsible thing to do for my family.

I questioned how I could remain a stalwart activist and selfless provider, to bridge the gap between these seemingly dissonant identities. And I couldn't understand why, when everything around me appeared safe in the ivory tower, I felt so lost and alone.

For a brief period after successfully defending my dissertation, I seemingly had everything an ambitious young woman like me could want: an advanced degree, a husband, a newborn child, and great job prospects. Despite the number of times self-doubt and inferiority had riddled my PhD experience, I knew that my department at Columbia didn't graduate students who weren't worth their salt. So when I submitted applications and had interviews for various scientist positions, I entered the rooms with confidence and a sense of belonging. My sights were set on the coveted Conservation Science Research and Teaching Postdoctoral Fellowship at AMNH, but I knew I had other options if I didn't get it. When the offer came in, I was honored and grateful, and also empowered to believe I was qualified for the position and would contribute a lot to the museum.

For a little while, it wasn't so bad. Things felt stable, secure. But as the routine made itself harder to ignore, I struggled to find myself amid the domesticity. The monotony of a nine-to-five was dramatically different from everything I'd done previously and everything I aspired to do. I had an identity crisis. If I wasn't traveling or studying animals, who was I?

My frustrations about my inability to experience domestic bliss carried into my workplace. At the museum, some of my male colleagues took their kids into the field. That wasn't an option for me, not only because I was breastfeeding but also because I was studying carnivores—big-ass bears. For my colleagues studying insects or leeches, it was easy for their children to tag along without the threat of danger or the inopportune need to nurse their babies.

I was genuinely happy to be a mother. But I can't deny that navigating my various responsibilities was challenging. I had a newborn baby, a new job, a new degree, and a new commitment to my partner. I didn't know how to handle all the stress and external pressure that came with caring for another human being and spouse while pursuing a demanding career in science.

Plagued by these discordant feelings, I lost myself. Each morning as I looked in the mirror, I was unsettled by the woman looking back at me—this settled, boring, nine-to-five mom doing wildlife work from the confines of a cubicle. Meanwhile, Oba couldn't have been more pleased. Dinner was on the table, planes were on the tarmac, and I was consistently at home each evening after work. He found happiness in my sedentary state, while I found contempt.

For the first ten months, I breastfed Zuri. And for the first year of her life, I didn't travel. As soon as she turned one, though, my itch was so intense that I said yes to the first opportunity that popped up: presenting at a retreat centered on diversity in the natural sciences. Travel was a legitimate expectation of my AMNH job, so it had always been more a matter of *when* I'd travel, not *if.* Even so, this trip was technically optional. But in the interest of persuading Oba, I framed it as a professional duty.

My mom agreed to come help with Zuri, which began a trend of her coming to help during all of my travel. Her ability to be a supportive mom and an amazing grandmother has allowed me to spend more time in the wilderness than otherwise would be possible. Shortly after weaning my daughter off of nursing, I boarded a plane to Colorado, secure in the knowledge that my daughter was being well cared for.

This trip would satisfy my nearly year-and-a-half-long wanderlust. I felt detached from Zuri, but in a healthy way. My new-parent anxiety was lower than ever. I realized that when I was close to Zuri, such as when I was at work twenty minutes away at AMNH, I was preoccupied with how she was doing. Yet across the country, there was nothing I could do if she missed me. This was freeing. I could be more present and content in the moment.

The center, located in the Rockies, was gorgeous. However, as I drove away from the Front Range and traversed deeper through the mountain passes, my breasts began to ache. They were filling with milk, which I hadn't anticipated. I was under the impression that I'd successfully weaned Zuri off nursing and had thus ceased producing milk. I started to fall ill, and I knew I could become quite sick if I didn't work fast.

When I arrived at the mountain cabin I'd rented for the week, I found that there was neither cell service nor internet connection, which meant I couldn't google what to do or call a friend for advice. In desperation, I spent hours milking myself over the sink, squeezing my breasts over and over again until milk began to spray out and the pressure and pain began to subside.

For the first time since leaving home, I began to cry.

Who was this woman hunched over a sink, milking her own breasts in the middle of the mountains? What business did I have presenting at a conference for badass ecologists? Why was I half a country away from my family? I was a nobody.

My husband was right. I was better off—safer, healthier—staying home. Even my body agreed. And I missed my baby girl.

AFTER THAT COLORADO TRIP, I got into the habit of distinguishing between my wild life and my "real life." My real life was when I was in New York: being a wife and a mother at home, going to my office during the day, and coming home to make dinner at night for my family. But every couple of months, I'd have another adventure. For a long time, I felt like I had to be apologetic about the field science, primarily with Oba, because my travel placed a hefty single-parenting load on him. It was optional at the level I'd reached in my career, since I was now a proper, working scientist. The pressure to stake my claim in one lifestyle over the other led me to cut back on my trips, especially the international ones. I also considered whether the real life I had been acting out should become, well, *real*. But this felt like settling, like compromise. And more than anything, it felt like I was being forced into changing who I really was.

In contrast to the impact my trips to the field had on my relationship with Zuri, they didn't seem to make a huge difference to my relationship with Oba. Our marriage was strong structurally, but I was beginning to feel that it was weak functionally. We were good, kind, smart, fun people who had our acts together and had wonderful family and friends. We'd fallen in love quickly, gotten engaged a year later, married a year later, and gotten pregnant a year later. On the surface, we were moving right along—the picture of Black excellence and achievement. In reality, our union was a void. We certainly weren't miserable, but we weren't connected.

Year after year, Oba's disapproval of my lifestyle, my interests, and even my career planted seeds of resentment within me. He never prevented me from doing what I loved, but he was up-front about his preference for me to transition into something more traditional. I felt trapped, and both of us were to blame.

For years, I had been so head-down and focused on my professional journey that I hadn't realized how critical it is for two people who love each other to check in—openly, honestly, and frequently—about how the journey in love is going. We had skipped many of those conversations, which led us to where we currently found ourselves. I wanted to make both him and myself happy, but those two desires were at odds. Similar to the bears and lions I'd been studying, I knew I wasn't meant to live in a cage, but I had somehow made a cage my home.

I had two choices in front of me: a life of wild adventure or a life of unfulfilling stability. I felt I was being pulled in two separate directions. As women in America—particularly those of us who are Black millennial women—we're not taught that you can have two amazing choices in front of you and yet desire a third. We're not taught to take the invisible path. And we're certainly never taught to make things up as we go—to be guided each step of our lives by our instincts and by our hearts.

In the animal kingdom, when a creature feels trapped, it's fight-or-flight. I will always wish I had chosen to fight. I should have fought for the fullness of my desires and for my passions to be respected and welcomed. I should have sought a compromise that truly felt right for both of us. I should have been brave enough to ring the alarm for a serious evaluation of our marriage, prompted by the difference of opinion about acceptable lifestyles for a wife and a mother. But I didn't. Instead, I chose flight.

Chapter 9

In early 2016, my days were consumed with working full-time at the American Museum of Natural History, playing the role of dutiful wife, and trying to figure out motherhood. These three strands of my life should have felt woven together in a way that infused me with strength and confidence, yet I was starting to feel more frayed than ever.

While I was struggling to navigate a reality that felt limiting to me, halfway around the world, an amazing discovery set in motion a chain of events that would change the trajectory of my life forever.

A young Malagasy college student named Leona was finishing her undergraduate program in ecotourism and initiating her final research project. She'd travel to some remote villages with a minimal tourism footprint to gauge their potential for and community members' interest in ecotourism. After these conversations and assessments, she wanted to map the main areas of Madagascar that were prime for ecotourism. A key motivation for Leona's project was her frustration with the way Madagascar's ecotourism had been historically Western-dominated. She wanted her work to help put power back into the hands of local Malagasy people.

In one of the villages she visited, folks told her about a nearby rainforest. Leona knew the area well. It was a primarily dry area with abundant grassland, used for livestock and some agriculture.

Stories of a hidden rainforest sounded like the residual rumblings of an old myth, especially since no one she spoke to had ever been to the location because it was so high in the mountains. But the people assured her it was there and hinted that it was full of biodiversity—including exciting wildlife species.

By the spring of 2016, Leona had completed her research project and was on her way to graduate. But she couldn't shake this feeling that perhaps the high-elevation rainforest existed—something that could change the course of Madagascar's natural history. So she sought advice from someone trustworthy and well-connected, Dr. Patricia Wright.

Dr. Wright, an American who is a world-famous lemur biologist and conservation scientist, splits her time between Stony Brook University in New York and her research center, Centre ValBio, located in Ranomafana National Park in the heart of Madagascar. As a biodiversity expert and lemur specialist, Dr. Wright has a robust history of helping turn Madagascar's rainforests into protected areas. She was also recognized for employing and empowering Malagasy scientists and community members to design conservation projects and programs that best fit local needs and interests. As a woman scientist in her seventies, she'd had her fair share of experiences breaking glass ceilings and was always excited to mentor and assist young Malagasy women.

Leona approached Dr. Wright with the idea that this secret, unexplored rainforest might be teeming with some of Madagascar's most endangered species and could become a major site of scientific discovery and a renowned protected area. Would Dr. Wright help her assemble a team to scout it out?

Dr. Wright had extensive knowledge of Madagascar's terrain and didn't believe this rainforest existed as described. She pulled up satellite imagery of the area and pointed to the maps, explaining that nothing indicated an expansive forest. The only thing the satellites detected was a long, thin strip of green, not nearly large enough to function as an ecosystem. Even so, the evidence was compelling, and both Leona and Dr. Wright knew that community members' words often held more credibility than what technology could show. Within a couple of months, the two women were strategizing a plan.

As part of their scouting team, Dr. Wright and Leona thought it would be useful to include photographers to visually document some of the preliminary findings. Dr. Wright knew a graduate student back in the States who studied orchids and pollinators in the tropics. He was experienced with field operations and expedition planning and was a talented nature photographer who'd worked with many prestigious conservation groups across the world. His name was Logan.

I'd been friends with Logan since 2013. We met at a conservation-biology conference through a mutual friend. My friend and I were helping to lead a diversity, equity, and inclusion (DEI) workshop at the conference, addressing the needs of a field that, then, was extremely white-dominated. Since my friend and I were both graduate students, we knew of a couple other Black female grad students attending the conference and pulled them into the DEI discussion. We ended up with a group of about four or five young Black women; some faculty mentors, including my advisor, Dr. Eleanor Sterling; a couple of Asian scientists; and one stocky white guy who looked like he'd wandered in from filming a nature show: messy hair pulled into a ponytail, wiry beard, weathered skin.

Logan had sought out the DEI meeting, and when my friend saw him sitting in the circle, she introduced him to everyone.

He and I struck up a conversation, and I learned that several years ago, when they were both undergraduates, they'd done fieldwork together on the Indonesian island of Borneo, studying orangutans. I immediately liked Logan, and he immediately liked me. My friend was a good judge of character, and her stamp of approval made me more comfortable in befriending him.

Logan's flirtatiousness was disguised as charm. Throughout his career, he'd managed to win over senior faculty and young students alike, and he particularly focused on service to Black women. In the DEI meeting, he volunteered to handle some of the more mundane tasks that were necessary to keep the newly formed DEI committee humming along.

Later at the conference, I sat alone near the back of the auditorium during the keynote speech. Logan slipped in next to me and we began to chat quietly.

At one point the keynote speaker suggested that we, as scientists, owe the world our knowledge and should use social media to share it. The speaker said we should all create Twitter handles, right then and there. For whatever reason, I'd been opposed to being on Twitter. I think I worried that it would be an addicting time suck. But Logan grabbed my phone from my hands, downloaded the Twitter app, and helped me choose a handle.

"I guess my Twitter handle should include my husband's last name, Davis," I said, casually mentioning Oba's surname, which I'd decided not to take. "Maybe it could be a little homage to him."

"That's a nice gesture," he said. "I'm married too—and my wife is Black." His proclamation was a little sudden, but I'm sure he was trying to express that he was intimately familiar with supporting someone moving through the world as a Black woman— that he was a true ally to her and could be one for me too.

At the conference and in the days ahead, Logan and I became fast friends. We didn't see each other much because he was attending graduate school in the South, and I was in New York. Not to mention that my fieldwork took me to Tahoe, and his kept him mostly in Asia. But through the DEI committee's work, the occasional conference meetup, and the periodic email, we kept in touch. Logan always wanted to know what I was working on and celebrated every victory I experienced along the way, no matter how small. He became the sounding board and cheerleader I lacked at home, and the support he offered made me want to grow closer.

One day, Logan got the call from Dr. Wright about scouting the rainforest with Leona and a few guides. He traveled to Madagascar and, after the group's expedition, they reported back that it was, in fact, dense, biodiverse, and potentially home to numerous endangered species. Logan was then put in charge of recommending some team members to further explore the ecosystem.

In mid-2016 he called me and asked whether I'd be interested in bringing my mammal-tranquilizing skills to Madagascar for a rapid biodiversity survey. He shared the project's backstory and told me that Dr. Wright was sourcing scientists of all types from around the world to come together for about a month and learn everything we could about the rainforest.

One of the survey's goals would be to determine which lemur species lived in the rainforest—and, in particular, whether any ring-tailed lemurs resided there. If they found ring-tailed lemurs, they would need to tranquilize one in order to get a DNA sample.

I laughed. Yes, I was partially qualified to do the job; I'd been trained to dart and sedate black bears and had been doing this

with teams for almost six years. But I'd never worked in the tropics or with primates, though I had a basic understanding of their
ecology. I'd also helped dart bears while they were up in trees—as
the lemurs would be—and catch them as they sleepily fell toward
the ground. And by 2016, I had acquired my PhD and was thus
considered a bona fide scientist.

I admit, I wasn't solely excited about the prospect of another
adventure. After three years of flirting at conferences, impressing
me with his DEI commitments, and offering unaccepted invitations to join each other in the field, this was a chance for us to
spend time together and see what happens.

As eager as I might have been to say yes, I couldn't do so right
then. I wasn't free like I'd been in 2013, when Logan and I had
first started talking about working together in the field. Zuri
was only one year old. I was a year into my postdoctoral fellowship at AMNH, which was a big deal with a lot of deliverables to
produce. I hadn't been on an international field expedition since
before I'd found out I was pregnant. This indicated that I'd apparently accepted a more settled-down life.

Yet I wasn't ready to say no. I left Logan with a maybe and told
him I'd get back to him.

The next day he called me again and explained that it was
crucial for the whole project to move forward right away. Word
about the rainforest had gotten out, and they were worried
that the logging industry would start using the rainforest for
resources—something that had happened many times before and
caused a lot of destruction in Madagascar. The goal was to secure
protected status for the area as soon as possible.

There's nothing like an urgent conservation mission to steel
my resolve. Upon hearing the major impact this expedition could
have, I went to work ready to make a big ask: five weeks off from

AMNH to join an international team of scientists and Malagasy leaders to find and categorize all the biodiversity in a newly discovered rainforest and use that information to advocate for its protection from the logging industry.

Thus far in my postdoc, I hadn't asked for any special accommodations. I'd heard repeatedly that the postdoc period is ideal for early-career scientists to consider side projects that allow them to branch out and diversify their experience. What better way to do that than to seize my first opportunity to work with primates in the tropics—alongside the legendary Dr. Patricia Wright, no less?

My supervisors were more surprised than excited. Typically, this kind of opportunity was earned after submitting an application and being evaluated alongside dozens of other highly qualified individuals. But this once-in-a-lifetime opportunity had come from a phone call from a friend. Since they knew it would be ridiculous to deny me something that would also make our institution look good, they said yes.

I didn't, however, know what Oba would say. Throughout our marriage, I'd found that my ideas typically were better received when I presented a fully formed plan to him rather than a fledgling idea. Perhaps it was the lawyer in him—he needed people to build a strong case to persuade him.

Similar to my work situation, I hadn't called in a big ask from Oba regarding either my travel or childcare for Zuri. I'd gone to Tahoe for six days at the beginning of the summer, and with our part-time nanny and my mom's help, that had gone smoothly. But this was a whole different situation: five weeks fully in the field with zero communication, on an island in the Indian Ocean, with potential danger. My fieldwork always had an element of danger, whether I was in the US or abroad, and

Oba always made sure to emphasize that. "You're a mother, Rae! It's not right for you to be putting yourself in dangerous situations!" he would scold me.

But I wanted this. And the professional reasons were second, maybe third, to the personal reasons.

I longed to transform back into the Rae trapped only in memory. The Rae who was part of African wildlife conservation and had epic adventures.

The pendulum of my life swung between two men: my husband, who would shake his head at the thought of me finding lemurs in a secret rainforest, and Logan, who would celebrate me for the same thing. The idea of a man being interested in me, giving me attention and encouraging my identity as a strong, capable wildlife ecologist, excited me. Perhaps most of all, I wanted to be seen.

I called Oba during the workday and gave him a heads-up that I had a fieldwork invitation that would be huge for my career growth. I said I wanted to discuss it that evening after Zuri was tucked into bed. I felt like that was a fair way to start, by giving him a bit of mental prep time and hinting that I was bringing something complicated to the table.

As part of my preparation for this conversation, I'd already called my mom, told her the stakes of this mission, and asked whether she'd swoop in and help by living at the apartment for at least two weeks. She readily agreed. Oba's mom, Valinda, was eager to help the other weeks, if she could change her work schedule around. We still had a couple of months to plan, and I'd asked Dr. Wright for a portion of the grant money being used to fund the trip to apply to my childcare needs. Dr. Wright, a fellow mother, had agreed.

When I told Oba about the expedition, he shook his head as I'd predicted, and emphasized the length of time I'd be gone. "You're a mother, Rae—you can't just go to Madagascar for a month! Put Zuri first." This was both Oba's logical response and his attempt to employ what he thought would be a surefire deterrent that would force me to reconsider my plans.

Shame once again reared its ugly head. In many ways, I agreed with Oba. I had zero examples of mothers who willingly left their children for prolonged stretches of time to chase professional pursuits. No examples of mothers who'd had many prenatal adventures but even more postpartum adventures. My request felt unreasonable, yet I knew it was what I wanted. *It's okay to want things*, a quiet inner voice reminded me.

"I have a plan for Zuri. My mom is happy to help for a couple of weeks, and your mom said she'd do a couple of weeks too. Plus, at this point in my career, I have a choice of whether to put myself in a bear-biology envelope or expand my area of expertise. Doing a lemur study in the tropics could really expand my opportunities."

I don't know why I said all of that. Maybe I was once again "pleading my case" in the courtroom of our marriage. Oba wasn't concerned with my professional opportunities. I had high-level degrees from fancy schools and a postdoc from one of the most notable institutions in the world. My career prospects weren't wanting. But what Oba didn't, and perhaps couldn't, understand is that I didn't want a job: I wanted a life of adventure. I wanted freedom. I wanted to constantly be experiencing things that were new and challenging to me.

Despite any misgivings he may have had, Oba agreed. Or perhaps, he resigned. He seemed detached from the situation, as well as fairly apathetic about my wants. My desire to travel for my career wasn't a hill he was willing to die on.

In early December 2016, I kissed my sleeping toddler goodbye. I felt confident that I'd return to her safe and sound, but what I couldn't be sure of was whether my leaving would cause an irreparable rift between us.

As I kissed her I prayed that when we were reunited, it would be as if I'd never left.

I took a taxi to JFK Airport, loaded down with a huge waterproof duffel bag filled with only essentials, since I knew I'd be hauling my luggage the entire trip. In my backpack I carried antimalarial medication, lots of inhalers for my asthma, a journal, a book, and a DSLR camera to use when my iPhone battery inevitably died. My plane landed at Charles de Gaulle Airport outside of Paris, where I met up with Logan and his best friend, Alex. Together, we boarded our flight to Antananarivo, Madagascar's capital.

We were en route for a couple of days and arrived in Antananarivo after sunset. I couldn't get a real sense of the city, but from the windows of the taxi that drove us to our hotel, it looked comparable to the bustling African capitals I'd passed through on my way out to the field. If anything stood out, it was the people. Although Madagascar is considered part of Africa, the heritage of indigenous Malagasy people isn't from mainland Africa but from South Asia. The people had skin as brown as mine, yet their dark hair was straighter and less kinky, and many people's skin had a goldish-red undertone. I realized that unlike my experiences living and working in East Africa, here I would stand out both in the way I looked and the way I carried myself. I'd routinely found comfort in being Black in Africa. I appreciated being able to blend in, so my identity wasn't constantly in question.

I hoped I'd be mistaken for a mainland African instead of pegged as an American. There was a stigma that came with being an American doing conservation work abroad, and earning the trust of the local people could be challenging.

We had a two- or three-day-long trip south to the village at the base of the mountain range that held the rainforest. The length of the journey depended on road conditions and how much it had been raining as of late. That night would be our only one in Antananarivo. A couple of hotel rooms had been booked for Logan, Alex, and me so we could get some sleep before joining up with a few more people to start our drive the next morning.

When we arrived at the hotel, the front desk clerk informed us, in a combination of broken English and French, that he only had one room available; there must have been a mistake with the reservation. Tired and without other options, Alex, Logan, and I carted our many bags up the stairs and opened the door to a large room that had one king-size bed under mosquito netting, one couch, a large bathroom, and a balcony that overlooked a gorgeous patio and a swimming pool. The American dollar goes far in Madagascar, and even a hotel at what would be considered a basic American price offered a fairly luxurious experience.

"I've gotta go to sleep, but it's a huge bed, so I don't mind sharing it." I was subtly talking to Logan and fairly confident he was reading between the lines. "We'll be sleeping on the ground for the next five weeks, so everyone might as well get to sleep in a comfortable bed one last time," I added.

I spoke matter-of-factly, but I mentally panicked a bit as I realized I only knew how to be one way in the field: *free*. I didn't know the rule book for women who were on an expedition but in their real life were married with a baby.

Not surprisingly, Logan volunteered to share the bed.

I climbed into bed dressed in leggings and a sweatshirt, and the next thing I remember was waking up alone in the morning. A note from Logan and Alex said they were downstairs on the patio eating breakfast with a friend. I finally took a shower and re-dressed in the clothes I'd slept in. After connecting to the hotel Wi-Fi, I used my phone to email Oba, letting him know we'd safely arrived in Antananarivo. I also said this would likely be my last correspondence for the foreseeable future. I hit send then shut off my phone to preserve the battery, also shutting off my connection to my life back home.

IN OUR EXPEDITION TEAM, the different specialists—botanists, small-mammal biologists, herpetologists, geologists, archaeologist, and so forth—had specific roles. My job, along with a trio of Malagasy field assistants, was to get evidence of ring-tailed lemurs. They're about the size of a house cat, and their long, bushy tails have a distinct pattern of black and white stripes, like a raccoon tail.

I'd neither visited the tropics nor studied primates, but I'd tracked and trapped African lions and black bears plenty of times. Dr. Wright and Logan's faith in my abilities infused me with confidence, but I still had my moments of doubt.

On the outside, I tried to project calm, cool self-assuredness, but I was freaking out on the inside. I indulged in catastrophic thinking, envisioning scenarios where every scientist but me nailed their objectives. I even concocted a situation in my head where I was the reason the park didn't gain protected status.

All the organisms we sought were important, but the ring-tailed lemur was the most crucial. Ring-tailed lemurs are critically endangered, and some estimates suggest there are only two thousand of them left in the wild. Finding lemurs would be the main justification for giving the rainforest protected status. So for me, it was an either-or situation: either I'd find lemurs and save the forest, or I wouldn't find lemurs and the forest—along with the species it housed—would be lost to logging.

Three trucks packed full of researchers embarked on the two-day drive from Antananarivo to the village at the foothills of the mountain range where the rainforest was located. Logan and I rode in the second truck. For two full days we sat in the back seat with our bodies smashed together and nothing to do other than talk and flirt. We dozed on each other's shoulders and whispered and giggled. He kept his long-lens camera on his lap, so it wouldn't get damaged from bouncing around in the truck. We couldn't drive fast on the unpaved roads, and the leisurely pace offered him many chances to roll down the window and snap photos of the breathtaking landscape. Though I risked draining my phone battery, I took pictures too: rice paddies, children running alongside our trucks as we passed through villages, and Logan—the classic rugged white male photographer and explorer with a beard and a dirty shirt and a professional-grade camera, traveling through Africa.

Was I socialized into thinking this image, which I'd spent hours staring at on the TV during my childhood, was attractive? Maybe. I'd dated mostly Black guys and a handful of white guys. Yet none of them were outdoorsy. I had no idea if I was actually attracted to this kind of person.

It was true that a palpable, exciting sexual tension was building between Logan and me. After years of sustaining a long-distance friendship, we were finally so far away from our "real lives," side by side on a life-changing adventure. The logical side of my brain turned off, and the present, passionate part took over. For the first time in years, I felt alive.

We pulled into the village, which had one through street that was more like a rough trail of brick-red dirt. The rain had packed it down like clay, which made for an atypically smooth ride. The residents' huts were made of the same red clay and spread out in either direction. Young children, dogs, and curious elders poked their heads out of their huts to see what was happening, and after we parked our trucks, the community members walked down to greet us. In addition to their local language, most of them spoke Malagasy and were able to fully understand our Malagasy scientists' explanation of why we were there.

All of the travel and the jet lag had fully caught up with me, on top of my self-induced dehydration. After a decade-plus of doing fieldwork, I'd learned that my weak bladder could be a huge inconvenience on drives to the field. For the team's sake, it was easiest for me to forsake water to minimize bathroom breaks. The self-empowerment superwoman in me would never give this advice to someone else, yet at that point in my life, I often sacrificed aspects of my own wellness to please others.

We asked if there were folks, primarily men, who had some free time and the energy and stamina to help us carry our supplies from the village to where we'd be setting up camp right next to the rainforest, about a five-mile hike. Several people signed up, and they were paid for their assistance. Sadly, there's a long legacy of conservation scientists abusing power and forcing labor on community members. I wasn't in charge of this expedition, but I was relieved that the villagers were paid for their participation and that a lot of the scientific leadership for the expedition were Malagasy scientists.

About fifty people set off together, including the community members helping us transport some of our gear. And we hiked. And hiked . . . and hiked.

I hiked all the time for work and had become quite skilled, yet this was different because everything—the ground, the rocks, and the vegetation—was so *wet*. That's what has stayed with me the longest—not the humidity or length of the hike, or even the considerable elevation. I wore the same hiking boots I'd normally use for fieldwork, but I noticed other group members were wearing flip-flops because we were wading through so many streams, rivers, and muddy areas. Despite my awesome high-tech hiking boots, everything was slippery. I slid into the water so often that I soon was walking in soaking-wet boots. As I started to envy the team members wearing flip-flops, I saw that most of the men from the village were trekking barefoot.

I made it to the top of the grassy hill and looked around. One lone tree on this high-elevation grassland provided some shade. It was beautiful and photogenic, like it was created to give the hill an eye-drawing element.

The folks who had reached the destination ahead of me were putting down their baggage, so I figured this must be the campsite. I had set up camp in all kinds of places, but I knew a large group would need a particular kind of space: one that's as flat as possible, not exposed to excessive wind, and near flowing water that can be used for drinking, cooking, and bathing. As I stood by the lone tree, I noticed that the area seemed to be on an edge. I couldn't tell whether it was the edge of a cliff or an illusion from my vantage point, so I walked over to get a better look.

When I approached, I discovered that we were at a high elevation, and the edge was a steep cliff that dove into a narrow sliver of canyon. From a distance, it seemed that with a running start, you could leap from one side to the other. But if you fell, you'd plummet forever. At the cliff's edge, I began to see what had been described to the team over the past few months. The sides of the canyon weren't covered in a continuation of the grassland we stood on. Instead, there was a rainforest—an entire rainforest. It was exactly like it had appeared in the satellite pictures: a dark forest-green patch with a cloudy, white mist hovering over it.

Carefully peeking over the edge revealed more of the gorgeous and unique expanse. My body could feel how sharply the slope plunged into the unknown, dropping in elevation with miles of landscape. But from up high you'd never know it. It just looked like a skinny canyon tucked between dry, grassy mountain ranges.

After a preliminary sweep, I pitched my tent and set up my personal area. I had a small yellow tent with a rain flap and an additional tarp for double protection. I had next to nothing inside my tent because I'd only brought what I could carry and hike with. My most prized possessions were a few packages of beef jerky and some protein bars from home, neither of which I was certain would last long in the high humidity. Everything grew mold.

Other team members began building a "bathroom," which was a deep hole in the ground surrounded by bushes significantly downstream from where our camp was, to avoid any risk of contamination. One hole in the ground for fifty people to poop and pee into. Although it's a part of fieldwork that makes me cringe, it's normal. Nobody said this work was glamorous!

Dr. Wright's husband was part of this first stage of the expedition, and he decided to engineer a shower. Basically, it was a two-bucket pulley system that looped over a tree branch. One bucket had multiple holes poked into the bottom, so when it filled with water, it drained out like from a showerhead. We'd take turns hauling buckets of water from the stream to the shower area, where we'd leave them in the hot sun all day to heat up.

Then we'd dump the water into the bucket with the holes, hoist it over our heads, and experience the decadent familiarity of a warm shower, even if it only lasted about sixty seconds. The trick was to soap up ahead of time while you were dry, so all you had to do was rinse. Dr. Wright's husband even enclosed the area with a couple of tarps, so we could shower with a bit of privacy.

I was going to be muddy, sweaty, wet, or all three most of the time, but each night, I intended to completely rinse off and change into my one pair of pajamas right before I crawled into my sleeping bag. It was only a semi-effective routine, but I hoped it would offer me some semblance of comfortable familiarity in a place where everything felt new.

Another motivator for my nightly bathing ritual was Logan. As impossible as it would be to feel fresh and sexy in the middle of a rainforest on an expedition, I wanted some semblance of the confidence that comes with cleanliness.

If we'd been taking advantage of escaping our real lives to lean into an illicit flirtation during our travel here, arriving at what felt like the ends of the earth solidified our desire for each other. I had no doubt that fate had brought us to a part of the world so different from anywhere I'd ever been, and so unknown by anyone. My real life thousands of miles away faded into the background. I set up my tent, thinking it wasn't a question of whether Logan and I would be physical, but when.

Due to our proximity to the equator, sunset was at 6:00 p.m. every day. Our first day on-site was waning, and I was ravenously hungry. Rice was boiling in a cauldron over the firepit at the camp cooking site. I changed out of my sweat-stained clothes and put on the one pair of leggings that I'd brought, a zip-up fleece for evening, and my flip-flops and went down to get something to eat. I deliberately avoided Logan because I didn't want to get looped into anything social that first night. Amid my fatigue, I was realizing I had a major job to do, starting at the crack of dawn. I got the first plate of rice, which had a delicious sauce, made from dried fish and lots of salt, poured over it. I devoured the food, walked back to my tent, zipped it up, put on my headphones, and used my last bit of phone battery to listen to music until I dozed off.

THE NEXT MORNING, my body naturally woke up as the sun rose. I felt a flash of pride; I'd beaten my alarm by two minutes. I sat up and pushed myself to get moving. Before I fully slid out of my sleeping bag, I scanned my tent for snakes, picking up my hiking boots and opening my backpack to make sure nothing had slithered into a crevice overnight. I didn't spot anything out of the ordinary, so I dressed in my field clothes, laced up my hiking boots, and unzipped my tent, emerging into a lovely, cool dawn, only slightly humid and with a light scent of smoke wafting over from the campfire that had burned all night long, keeping water boiled and sterilized and ready to be used to cook early-morning breakfast.

Each team member had a different fieldwork schedule, but mealtime was the great equalizer because it would occur at the same time every day. You snooze, you lose. This particular day's breakfast was no different. I squatted on a fallen log and gratefully scooped heaps of plain boiled rice seasoned with salt into my mouth. It wasn't particularly delicious, but it was all we had, and I recognized it would power me through the first half of my day.

Logan had beat me to the cooking site, and he handed me a tin mug full of *ranovola*, a staple in traditional Malagasy cuisine. It's a tea of sorts, made from bringing water to a boil with the burned rice from the bottom of the pot used to make breakfast.

I like tea and found comfort in the hot drink with my breakfast. And Logan catering to me felt like more than him being polite.

This morning I'd begin working with my team, who were all Malagasy and highly experienced with lemur tracking. They were going to teach me everything they could in an abbreviated amount of time—and all without speaking the same language. They spoke Malagasy and French, and I spoke English, Spanish, and Swahili, so we used sign language, body language, pushes and pulls, and an overabundance of *oui* and *non* to communicate basic messages. But even without being able to fully communicate, I could tell that our morale was high.

Before I could finish my rice and ranovola, Velo, the man I'd describe as second-in-command of our group after me, yet who was the most experienced in lemur tracking, walked up to me and stood there. I knew exactly what this meant: he was non-verbally telling me, *You're late—let's go.* I handed my unfinished plate of rice to Logan, who was happy to have extra to eat. Then I threw my backpack on and followed Velo to meet up with the rest of the team, Joseph and Tsiry, who picked up their packs as Velo and I approached. They didn't have as much lemur-tracking experience as Velo but certainly had more than I did. Joseph would prove to be best able to hear the lemurs, and Tsiry was by far the fittest and fastest. None of us had ever done wildlife work in this particular rainforest, so the newness of the landscape and of the differently behaved lemurs united us.

Neither Joseph nor Tsiry were wearing shirts that morning. It wasn't yet seven o'clock, but it was already warm, and we were about to start the most physically intense part of our day.

The camp was situated on the side of a mountain, where wind and rain were the least aggressive and where we were close to a stream. But this placement also meant that in order to access the trail my team used each day for our lemur hunt, we had to scale a hill that was steeper than steep. Although I aimed to keep an open and positive mind, it was impossible not to think, *Let's get this over with.* Our journey started with a solid twenty minutes of huffing and puffing and sweating and willing myself to keep going.

I began my ascent and only one minute in, I was out of breath. I concentrated on the landscape to distract myself. Even in the daylight, it was dense and fairly dark, yet welcoming. It was fertile and bursting with life and full of sound. It smelled fresh, which was helpful for the strenuous hike, and the vegetation made it feel as though we were stepping back in time. The ferns and orchids growing on the side of the ancient trees reminded me of what dinosaurs must have walked through millions of years ago.

Finally, I made it to the top of the mountain, catching up to Velo, Joseph, and Tsiry. I tried to hide that I needed a second to catch my breath. In many ways the climb had been torture, but it also provided me with an initial sense of accomplishment. I only had a moment to rest because Velo, Joseph, and Tsiry were already heading into the forest, machetes in hand as they hacked their way in. Entering this part of the forest was sort of like entering a den; the density of plants made it dark. And unlike our campsite, this part of the forest was at a high elevation and loud with the sounds of rainforest birds. I couldn't see them as they soared above the tree canopy, but their high-pitched calls reverberated through the air.

We strained to hear the rustling of branches that signaled lemurs hopping from tree to tree. I held a naive hope that they'd simply be there, waiting for us. We walked farther down our path, listening for their calls, either a moan or the "hmmm" sound that makes up their special language. Lemurs live in groups of up to thirty, so if we heard anything, a bunch of them were close by.

My backpack weighed down my shoulders, even though I only carried the basics: water, some protein bars, a GPS unit, and a tranquilizer gun. The gun, which looked fairly intimidating, fired tiny darts full of sedatives that would anesthetize a lemur briefly enough for us to gather data before releasing it back into the wild. The GPS unit was indispensable because it would allow me to record the location every time we found evidence of lemur presence.

That first morning turned into the afternoon, and before I knew it, Velo suggested we return to camp. We were spent from hacking a path with the machetes, and we knew that if lemurs lived in the forest, they weren't at the edge. We'd need to spend days delving into the interior.

We retraced our footsteps back to camp. Most of the other scientists were still out setting up their data-collection points, but I saw Logan sitting with a team member, looking at maps and seemingly planning some logistics. When he saw me, his eyes lit up.

"We didn't find anything yet," I said before he could ask. I smiled and told him I was going to take a shower.

As I ducked into my tent to grab my towel and soap, I felt a pit in the base of my stomach. *Rae, are you about to seduce this guy?* How far was I going to take this?

I walked to the shower area in a daze, mentally rehashing all the times I'd flirted with men while I was with Oba. I'm not proud to admit that I'd been deliberately flirtatious plenty of times. However, what was happening with Logan was beyond flirtation: it was an invitation to physical intimacy.

The sexual energy and excitement I experienced with Logan highlighted the reality that those were lacking in my marriage. I longed to feel physically good and wanted and believed I deserved that. And here was an opportunity, with a person I knew well enough and trusted. I wasn't in love with Logan, and I didn't hold some romantic notion that he was my long-lost soulmate. Maybe it was adrenaline or my broken pieces guiding me irresponsibly away from my values, or both. We were already swept up in a whirlwind of adventure, so why not add another surge to the storm?

At sunset our camp transitioned from sweltering hot to dark and cool, and rain began to fall, extinguishing our fires and forcing us into our tents. Logan had a bottle of rum he'd purchased at the duty-free shop in the Paris airport, and he was offering a drink to anyone who wanted it. He raised his eyebrows when I declined. Even after all the years we'd known each other, he still had much to learn about me. I made a habit of taking impeccable care of myself when I was in the field. Excluding alcohol was my norm because we had no access to medical care, so I couldn't afford to get sick, even with a hangover. Plus, I have a famously low alcohol tolerance and tend to fall asleep after one drink. That night, I didn't want to miss a thing.

"I'm going to my tent, but it'll probably take me an hour or two to fall asleep," I whispered to Logan. I sashayed away from the campfire, as much as one can in grimy field clothes, leaving sexual tension in my wake.

About fifteen minutes passed before Logan's voice pierced through now-torrential rain. I unzipped my tent, found him standing there soaking wet, and welcomed him in. His battery-powered lantern provided excellent illumination and created a much more romantic atmosphere than my dim headlamp.

He removed his wet jacket, pants, and shoes and put them outside the entrance, where I'd designated a shielded space for wet, dirty items.

We were both cold, and the seasoned survivalists in us knew that we could warm ourselves with one another's body heat, snuggled in my sleeping bag. As we undressed, I needed to address the elephant in the tent.

"I really want this," I said while making firm eye contact and smiling. I couldn't remember ever having given verbal consent before kissing a man. Nor had I ever been asked. Yet here I was, taking the initiative to make it crystal clear that I desired this intimacy, that I was a willing and eager participant. Logan's response was a passionate kiss that started on my lips then migrated all over my body.

I recognized that my verbal consent not only gave him the green light to proceed but also freed me from doubt and confusion. After years of self-denial, I let my voice express my desires. I was open and free, both physically and emotionally.

There on the forest's edge, while a tropical rain pelted the tarp protecting my tent, two married people who were thousands of miles away from their spouses had delicious, explosive sex.

Logan woke before sunrise, kissed me, and snuck out of my tent so nobody would see him. I lay awake, recounting every moment. A part of me that I'd missed had been awakened.

I had officially cheated on my husband, and I'd eventually have to deal with what that meant. But right then, I gave myself permission to let the big feelings surface. And I felt amazing.

For the next two weeks, I woke up at dawn and spent each day carrying out fieldwork and each night with Logan. At times I couldn't believe I was being gifted with such a full experience in Madagascar.

My team hadn't found ring-tailed lemurs yet, but the other scientists were making discoveries. Also, I was viewing wildlife I'd only seen on television, learning new skills, making friends, and having profound discussions with the Malagasy team members about the legacy of colonialism.

On a personal level, I was experiencing physical intimacy and building a relationship with someone who felt like a great match. Before long, our guards came down, and Logan started leaving my tent even after other people had begun to awaken. Our lovers' bliss was apparent to everyone, but it seemed understood that whatever happened in Madagascar would stay in Madagascar.

While some parts of the expedition were magical, many parts demanded the toughness that adventures require. Sometimes I can't even begin to describe what it was like to live in that rainforest for five weeks. The physical challenges abounded, from the intense rain to our makeshift, temporary accommodations.

And I encountered something completely new because I'd never worked in the tropics: tree leeches. These leeches live on tree leaves in the rainforest canopy. When it rains, they slide off the leaves and drop down onto animals and people. I'd find leeches in my hair or on my neck, where they'd fallen without me even realizing. That's the cool thing about leeches—they start sucking your blood, and you have no idea.

One of our Puerto Rican colleagues had gone out for the day, and leeches had fallen onto his head, then gone down his neck and onto his back. When he returned to camp, he removed his pack, and the back of his shirt was a bloody mess. He was completely oblivious. We made him take off his shirt, and we all took turns peeling leeches off his back.

Another complication revolved around the lack of methods to deal with menstruation. For whatever reason, my body kind of revolted against me, and I had my period during the whole trip. Weeks of menstruation were certainly not part of my plan. I'd done expeditions where I was the only woman, and I'd done lots of trips to the field alone. But this expedition had a maximum of four women at a time, with forty-plus men, and sometimes I was the only woman at camp.

I didn't have enough products to tend to my needs, so I resorted to my survivalist skills and attempted to create menstrual products from natural materials like mosses and grasses. I collected some moss, dried it in the sun, and shoved it in my underwear. It temporarily worked but ultimately was a mess, and an inconvenience among everything I had on my plate.

Two weeks into the trip, I was torn between wanting the expedition to be done because of the huge physical burden and wanting it to last forever because of the affair with Logan.

I also felt like a failure because I still hadn't found and captured a lemur. After weeks of hard work, I had nothing to show for my effort. We'd traversed the entire rainforest, and although we could hear lemur calls, we hadn't spotted any. Time was running out, and my reputation was on the line.

About three weeks into being tired and stressed, my team hiked into the rainforest, along the same path we'd started clearing out our first morning. The path was less dense thanks to the previous weeks of hacking. When we got far enough into the rainforest, we stopped and listened—and heard the familiar ruckus of ring-tailed lemurs. Through the thickness of the trees, we couldn't see them clearly, just blurry movement and rustling branches no more than fifty feet ahead of us. It was time to chase these primates down.

They screeched their alarm calls to each other as we ran after the pack of lemurs that bolted away from us, bounding from tree to tree. We desperately needed to keep track of them, so we could try to locate them again the next day. But we couldn't keep up and eventually lost them in the trees.

Yet a few moments later, I heard a lemur call and my eyes locked with Velo's. *"Maky!"* I said, using the Malagasy word for "lemur." He started listening and heard it too. Immediately, he stood and started running toward the sound.

We saw a couple of lemurs, one perched on a relatively low branch, as if it were inviting me to capture it. I needed to grab my tranquilizer-dart gun, but the lemur captivated me. After weeks of pursuing these creatures, I was finally seeing one in the flesh.

Velo shoved my shoulder. "Dart it!" he said.

I snapped out of my stupor. This might be the best shot we'd ever have. I unstrapped the dart gun from my backpack and quietly loaded it with the tiny darts full of sedative. I looked through the target ring on top of the dart gun and aimed at the lemur. I took the shot.

The bright-pink dart soared through the air then fell back to the ground. The lemur, sensing imminent danger, started running away.

I'd both missed the shot and lost the dart, two embarrassing screwups we couldn't afford.

But my team wasn't ready to give up. They took off running, trying to keep up with this lemur family that seemed more willing to let us get close. I caught up and found Velo crouched behind a bush. He pointed to a low branch where what appeared to be that same lemur was sitting, grooming itself.

I got my gun out and aimed it so the dart would hit the lemur's hip. Then I pressed the trigger and took the shot.

It hit!

I stood up and thrust my arms in the air, silently cheering for myself.

As thrilled as I was, we didn't have time to celebrate. We had to keep hiding and make sure the lemur didn't run away. Miraculously, it stayed put while it looked around and turned this way and that.

Velo pulled a folded-up tarp from his backpack. He handed me one side then motioned for Joseph and Tsiry to come over. Without speaking, we each took a corner of the tarp, preparing to catch the lemur as it fell from its perch, an exercise I'd done in Tahoe with bears numerous times.

We could see the lemur's eyes slowly blinking, the same way humans' eyes do when we're trying to stay awake. We positioned ourselves under the tree, and because the lemur was only a couple of meters above us, I got a good look at it. Its back looked extra big and fluffy, and I realized: *it had a baby.*

This lemur was a mother. Which meant this conspiracy of lemurs was a breeding population. As a breeding population, it was increasing in size, which was tremendous information and possibly spelled out a new future for ring-tailed lemurs in Madagascar.

Suddenly, the baby hopped off her back. We saw that it wasn't a vulnerable newborn. It was a slightly older baby, mature enough to be on its own and not necessarily attached to its mother.

The mother fell asleep and dropped from her branch, and we gently caught her in the tarp. We put her on the ground and double-checked that she was okay. Joseph, who had the heaviest backpack full of gear, pulled out a soft towel to wrap her in. How we'd managed to preserve a clean towel in a damp, muddy rainforest, I have no clue.

After weeks of living in a secret rainforest in the middle of Madagascar, with a team of young men who had become close friends, achieving something so monumental so fast gave me whiplash. I'd played out this scenario in my mind many times, and I'd even journaled about my doubts and uncertainties. But as I paused to admire her, stunned and awestruck, I noticed the humanlike details of her tiny face, and intense emotions stirred within me.

Just like me, she was a mama with a baby mature enough to be detached for short periods. Just like me, she was wild and forging a life in an entirely unexpected place. Just like me, she was doing her job and doing it well.

Velo once again had to grab my shoulder and nudge me back into the urgency of the moment. His eyes flashed as he yelled, "Rae, go!" and pointed back toward camp, where an animal-processing area had been set up for the entire expedition team. Any captured animal was brought to the camp, where it was given a checkup, sampled for data collection, and photographed. We then returned it to the rainforest and released it.

I came back into my body and remembered that while capturing a lemur was the most essential part of the process, my main job was to get her to the camp, collect the data, and return her to the site where she was found, to reunite her with her baby. Because she was small and nursing a baby, we didn't want to administer additional sedatives. This meant we had about one hour, max, before the sedatives wore off.

My team needed to stay at the site where we caught the lemur, to keep an eye on the baby and make sure we knew exactly where to rerelease her. For the first time during the whole expedition, I had to make my way through the rainforest, all the way back to camp, by myself. If I'd had more time to think clearly, I would've excused myself from this part of the process. It was way too dangerous for me, still unfamiliar with the vast, dense rainforest, to navigate back to camp on instinct.

It had never dawned on me that this high-energy, high-stakes part of the mission would come after I was already completely exhausted from a long day of running around the rainforest trying to catch these elusive primates.

Joseph thrust the wrapped-up lemur into my hands, and Velo yelled, *"Go!"*

These amazing Malagasy men who'd spent much of their adult life in rainforests like this one knew that although they were critical to the mission, the lemur capture was my big win. They wanted me to bring this mission on home.

With the lemur mama bundled like a baby in my arms, I ran through the rainforest, up the incline of the canyon. I was going in the general direction that Velo had pointed me in, hoping I'd soon see a recognizable landmark.

This was my hero moment, and I wish I could say I was being my most powerful, most mentally strong self. Instead, anxieties bubbled up within me, and I started panicking about the load of living, breathing responsibility I held.

But I didn't stop running. In the same way I had on the uphill hikes that had felt impossible during the first week of the expedition, I put one foot in front of the other, forcing my body forward while my mind was paralyzed. I gave myself the mother of all pep talks, out loud for the entire rainforest to hear: "Rae, you've got to do this. You're going to do a great job. You're going to process her. It's going to be fine. She'll be safe and healthy. You'll find your way there. You'll find your way back."

I tried to predict what the next few minutes would entail. I held a wild jungle primate in my hands—which also happened to be one of the most critically endangered species on planet Earth. And I was responsible for collecting what could be life-changing data.

Until then, I'd never been the "lead" scientist in any capacity, on anything. Sure, I'd done my own independent research, but my first time touching a lion or a bear was always in the presence of a seasoned scientist who showed me what to do. This time, I was the senior scientist. I'd left my team back in the forest, and the other mammal experts were elsewhere doing fieldwork. So I'd have to go through the entire process on my own. One tiny slipup, and all could be lost. Processing the lemur included straightforward data collection like weighing her, measuring her body, and evaluating the condition of her eyes, ears, fur, and general health. But it also included taking a blood and a hair sample, putting a collar with an ID tag around her neck, and ensuring she remained healthy and cool under the anesthesia.

"It's just like a bear. It's just like a bear. You've done a million bears. It's just like a small bear," I chanted as I bolted through the trees.

A ring-tailed lemur is decidedly nothing like a small bear, and I'm sure any real primatologist would be horrified to learn I was going into this data collection as if I were in the mountains of Nevada with a black bear cub.

At last, I mounted the final hill and saw camp on the horizon. I arrived breathless, unable to see anyone. I screamed, *"Maky! Maky! I have a lemur!"* For three weeks, everyone had been hoping to hear these words. Yet every afternoon, amid the cacophony of all the other rainforest animals, my team and I walked back into camp in resigned silence, empty-handed, shoulders slumped.

After the first week, the team had even stopped leaving out the processing equipment.

A Malagasy scientist named Herman was the first to hear me, and he stumbled out of his tent, eyes wide. He didn't say a word as he began laying a tarp on the ground, along with a bunch of different blankets. He then pulled out a huge tackle box full of the materials and tools we'd need to process the lemur. It had been three or four minutes since I'd collapsed on my knees in the middle of the camp, in complete disbelief that I'd made it there.

Herman and I still hadn't spoken to each other, and I was having difficulty catching my breath. My asthma was now in full swing; the adrenaline had made me unaware that an asthma attack was building up. After the tarp and the blankets were on the ground, I laid the lemur on her back and unwrapped her. The first thing I did was take a pair of tiny scissors, snip a bit of hair from her tail, and put it in a plastic bag, so it could be used for a DNA sample. I had Herman label the bag "ring-tailed lemur, female, number one."

In the whirlwind of activity, I hadn't noticed Logan come out of his tent, camera in hand, in full-on photo-documentation mode. I didn't hear the camera's constant clicking and whirring, but Logan was capturing everything.

I had wanted this moment for many legitimate conservation and professional reasons. But I also wanted to solidify the approval of the guy who had brought me here. I had wanted to look sexier and more badass, for me to be viewed as the beautiful, determined Queen of the Jungle that I longed to be.

When I look at the photos now, it's as though I'm transported back in time: I am drenched in so much sweat that my gray T-shirt looks soaking wet. I am focused and intense, looking at the lemur, the tools, Herman—everywhere but at the camera.

This event was momentous—not only to me but to the whole conservation community and perhaps the world. One of the expedition's main goals was to prove that ring-tailed lemurs lived in this rainforest, and the best way to prove that was with a combination of photos and DNA. And there I was, with a lemur in my hands.

I asked Herman for a syringe, and I found a large vein in the lemur's thigh to take a blood sample. I then weighed her on a hanging scale—a whopping four pounds—and called out her body measurements as Herman jotted them down. The last step was to slip a neon-green fabric collar on her. Down the road, if any scientist saw a ring-tailed lemur with a neon-green collar, they'd know it was the one we'd found. As we finished processing the mama lemur, her eyes began to blink, and her limbs began to stir.

For the first time since I'd arrived in camp with her, I looked around. Everyone present at camp surrounded Herman and me. Many of us were seeing a wild ring-tailed lemur for the first time, and we were mesmerized.

I badly needed my inhaler, so I grabbed it out of my backpack and turned away from my colleagues, a bit embarrassed of my need for it, and took a couple of puffs.

I wanted this part of the adventure to end. I wanted success to be sealed then and there, and I wanted to lie down and process the magnitude of it all. But there was a final, monumental step I once again had to accomplish on my own: I had to return the lemur to where we'd found her, and I had to run as quickly as I had on the way here a mere twenty minutes earlier. I didn't even remember exactly how to get back to my team.

I was exhausted. I was satisfied and I was ready for a break from the intensity that was my life. I wanted to hand off the last part of this job to someone else. I didn't want to face a final task with such a high possibility of failure. Yet if I didn't attack this task with competence and precision, neither the lemur mama nor I would be okay.

Nothing was going to stop me from getting the mama back to her baby before she fully woke up. I carefully cradled the lemur and ran. I was running in the direction from which I'd come, but much more purposefully. As I ran, I realized that my sense of direction was working for me. I didn't know exactly where I was going, but I was being guided by an instinct, and I felt confident I was headed the correct way. A few times I stopped to decide whether to go forward, left, or right, looking for signs of where I'd been before.

I still have a couple of scars on my chest from where branches scratched me so badly I bled. I didn't care about my body or what was happening with me. I had to end this mission with a win.

My team must have heard me before I knew where they were because Velo whistled to signal, *Hey, we're over here, come this way!* It was loud enough to catch my attention but not so loud that it startled the baby, which was still waiting for its mama. That whistle helped anchor my navigation. I arrived wide-eyed, stunned, and breathless. I didn't have the bandwidth to communicate in broken Malagasy, so I gave them a thumbs-up to tell them she was okay.

I knelt at Velo's feet, unwrapped the lemur, and laid her at the base of the same tree where she'd first fallen into our arms. I put a few drops of water into her mouth and backed away. My team and I hid in the nearest bush, not making a sound, and watched. The mother lemur woke up, looked around, and climbed up the tree to her baby, who latched on to nurse.

Tears welled up in my eyes. I, too, would be reunited with my own little one soon enough. In the meantime, I congratulated myself on having achieved what might have been the pinnacle milestone in my career.

After the baby returned to the mama's back, she darted off into the forest, out of sight. My eyes stayed on the branch where I'd last seen them, and I pondered the vital role I'd just played in scientific discovery.

That night, back at the camp, while eating rice and sitting beside the campfire's warm glow, Velo, Joseph, Tsiry, and I excitedly shared the day's events with the wider team. As we rambled on in frenzied animation, Logan looked at me with pride and tightly wrapped his arm around my shoulder.

I wanted to fully join the fun, but I needed a brief respite. We were all celebrating a shared accomplishment, but I couldn't ignore the depth of my own feelings. So I excused myself and walked to my tent.

I kept thinking about how I was completely different from every other person around that campfire. I was one of only a few women in the group, and I was the only mother. I was the only Black American. I was the only person married to a disapproving husband. As far as I knew, I was the only person with an insecure financial background and the only one with secret ambitions to become a household name based on my work in nature. And I believed I was the only one struggling with self-esteem, self-honesty, and the guilt that came with living my life the way I wanted to live it. I needed space to congratulate myself, by myself—from me to me.

I pulled out my journal and wrote a short reflection to capture the buzz of emotion swelling in my body:

*I feel like such a badass. Like, I am the shit. I am so cool and
unstoppable. I'm doing something new and adventurous and
it's awesome. And it's for the betterment of the planet. Today
was a good field day :) Before I finish, I want to express my
gratitude to the universe for today's success. For this
opportunity and for the upward momentum of my life. I'm so
amazingly fortunate.*

After writing in my journal, I was ready to rejoin the campfire
and receive the other validation that I sought: admiration from
Logan. I sat next to him and placed my hand on his knee, giving
it a squeeze. I locked eyes with him, mouthed the words *I'm so
happy*, and allowed a genuine, irresistible smile to brighten my
face, my lips intimately close to his.

Many of the expedition crew circled the campfire, including
Leona. She was with us during the whole expedition, but I didn't
interact with her much because I spent most of my time tracking
the lemurs.

At the sight of Logan and me, she turned away. Like everyone
else at camp, Leona undoubtedly was aware that Logan and I
were hooking up. People knew we both had spouses back home,
but no one said anything to us.

I dismissed Leona's funny look. Maybe she and Logan had
flirted when he'd previously visited the site. He did have that
tendency, after all. And perhaps she was jealous that his attention
was now directed elsewhere.

But I was unapologetic about our relationship. And I had no
regrets.

The next day started and ended much like any other. My team
and I headed out on another lemur mission. And found nothing.

We never saw the mama lemur again, or her group. We heard many more lemur calls, always a short distance ahead of us, teasing us like they had for weeks before. But we never captured another one.

At the end of the expedition, I traveled from Madagascar back to New York, with new muscles—both physical and mental—new field experience, new stories to tell. And a new love that had sprouted across the world but would follow me home.

SINCE I LEFT MADAGASCAR, I've heard a couple of tidbits of information about the mother lemur. A small team of graduate students braved the rainforest, setting up camp for several months to monitor the lemur population and try to better understand their behavior. Because of the mama lemur's neon-green collar, she's easy to identify. And I've heard she's doing great.

As for the secret rainforest, I wish I could say it's protected and has become a national park. These processes take many years to formalize and involve a lot of government cooperation, which is tough. But I'm happy to say that it's still intact and thriving with life. Most importantly, it hasn't been pillaged for resources, and it's well on its way to receiving protected-area status. It will become a beacon for conservation in Madagascar.

Looking back years later, it's obvious that capturing the mama lemur helped me to permanently see myself as an incredible, capable woman. Although I had countless reasons to feel that way about myself prior to that point, that experience completely altered the risks I'm willing to take, the value I believe I have, and the imposter syndrome I'm now able to tame. I came away from Madagascar knowing that I'd done something brand-new—something new not only for me but also for the world. Before that day, no one had ever caught a lemur in that rainforest.

I wish I could say that these feelings had always been within me, and maybe they once were. Perhaps little-girl Rae felt unstoppable, and over time that light was dimmed, as life experiences and society shaped me into the adult Rae who merely *projected* confidence. And perhaps my Madagascar success was a manifestation of "fake it till you make it." Regardless, I stepped off the plane in New York a different person, empowered and motivated to keep adventure as a prominent part of my life and my work.

This newfound inner strength coincided with a newfound affair, a relationship that had the potential to be glorious, authentic, and affirming. Little did I know the relationship that gave me so much during such a transformative time in my life would become manipulative, deceitful, and destructive in ways I never could have imagined.

Chapter 10

Madagascar's palms, ferns, and bamboo gave way to New York City's glass and brick. The humid air of the cloud forest was replaced with the stifling stench of last week's trash and yesterday's piss. Rats took the place of lemurs, and concrete with unidentifiable stains stole the place of the jungle's rich underbrush.

I arrived back in the States on Christmas Eve 2016, and on Christmas morning, we drove to my maternal grandmother's house in New Jersey to celebrate. The holiday festivities allowed me to avoid dealing with my inner turmoil, as I threw myself into the dual role of dutiful wife and mother of an eighteen-month-old daughter. I knew full well that at some point, I'd have to reckon with the new reality I'd created for myself.

Being home was both a relief and confusing. On one hand, I'd been looking for a way to escape my real life. However, part of that life included my favorite person in the whole world, Zuri. Additionally, I struggled to reconcile the person I had become in Madagascar with the Rae everyone had known prior to my trip. While abroad, I'd felt like a younger, freer version of myself—a version I recognized and was more comfortable being than this wholesome, stroller-pushing, buttoned-up mother in New York. It felt more natural to wake up to the calls of the wild than to my alarm's blaring call to action. A morning commute up the barely distinguishable paths through overgrown jungle brush was far more exciting than a ride on the downtown C train.

I missed the freedom Madagascar offered me. And I missed Logan even more.

Folding into Logan, I'd escaped the duties and commitments of my city life. I'd become accustomed to his physical affection and intimacy. Oba and I loved each other and were loving toward one another, but passion had waned, as is often the case with new parents. And Oba didn't seem very interested in hearing about my adventures. He didn't ask me many questions about my time in Madagascar. All he did was welcome me home with open arms, eager to reestablish the familiar comfort and routine that was his highest priority in life.

So perhaps even more important than the physical relationship I'd established with Logan were the intellectual and emotional connections he provided. His belief in both my abilities and my dreams touched a part of me that longed for affirmation. In New York, Oba was a successful lawyer, wining and dining with Manhattan's Black elite, while I sometimes felt like a doting accessory for him to tote along. I morphed into a housewife whose collar grew tighter and tighter with each corporate dinner and company party. In my angst, I pushed aside the idea that I was playing a role in my own captivity, by pretending that I was okay instead of communicating about my challenges.

With Logan, the possibilities seemed endless, perhaps too good to be true. Stories of his time abroad encouraged me to do the same. His unfaltering belief in my dreams to pursue conservation and ecology in the media made me feel seen. I felt smart. I felt uplifted. And I felt beautiful.

In many ways, I felt like Madagascar had given me a second chance to be the adult woman I wanted to be—to be the thirty-one-year-old I was on the inside. During the six years I'd been with Oba, to avoid conflict and criticism, I'd diminished who I was and how I wanted to express myself. The real me is a passionate, fearless-in-the-face-of-adventure, sexual, daring person.

Falling for Oba wasn't a bad decision—he's a great man, responsible and dependable. But his belief system was that marriage meant a different kind of relationship than when you're dating. In marriage, while he became predictable and happily distant, my desire for all of the wild parts of me grew more intense. I wasn't encouraged to define motherhood on my own terms, and as my vision for a wild life strayed further from the kind of wife and domestic bliss Oba desired, his support for my goals decreased.

On the plane ride home from Madagascar, I checked in with myself. In a moment of brutal honesty, I realized I didn't feel sorry for having sex with Logan. I did, however, feel *very* sorry for cheating on Oba, for the betrayal that I facilitated. But having passionate, new sex, and having a partner who was attentive and matched my desire, was something I felt that I deserved. The revelation fueled a powerful self-hatred. It was like my brain was split in half, with one side expressing honesty and the other side casting extreme judgment.

Even though self-condemnation had taken root within me, I decided to see if I could have the best of both worlds, to pretend I wasn't cheating while constantly trying to figure out when I'd see Logan again. We were rarely in the same place. We lived worlds apart, him in the South with his wife and me in New York with my husband and baby. But every couple of months we managed to rendezvous in the same city. I'd receive an invitation to give a presentation out of state and began saying yes to more travel opportunities. Zuri had become more accustomed to me being away from home, and in my absence, I knew she was receiving amazing care from Oba, my mom, and our village. These experiences allowed me to accomplish two objectives: advance my professional goals and reunite with Logan for one or two nights of steamy ecstasy. Thus, our affair continued through most of 2017.

Even so, as Logan and I grew closer, I began to recognize things I didn't understand. Stories began to misalign; facts began to falter. Questions flooded my mind as I learned more about Logan's mental and emotional state. As our relationship deepened, he began revealing some of the damning secrets hidden in his closet. He told me that he'd had experiences with painkillers and alcohol (he described them as past problems, but I wasn't so sure) and that he'd cheated on his wife several times. He also shared that he'd been in some serious physical fights while in the field. He said he didn't even remember the altercations well because they'd occurred when he was under the influence. For all he knew, they might have ended with people being seriously hurt or killed.

When Logan started opening up about these things, I allowed myself to be a safe support system. But my rational brain started building a case that Logan was a walking red flag. Some of the stories he told me about his life in the tropics were unbelievable, and on more than one occasion, I wondered if he was a pathological liar. Gradually, the details upon which our relationship had been built began to reveal themselves as pure fiction.

In the summer of 2017, a confluence of societal pain, professional disenchantment, and romantic rejection pushed me to seek out a new environment, albeit with an increasingly toxic person. I decided to pack up my two-year-old daughter and myself and fly down to Panama to track jaguars in the rainforest, an opportunity that Logan had pulled together with several of his colleagues from his previous tropical-biology expeditions.

In all the years I was a scientist before having Zuri, it wasn't abnormal for me to be in the field for as long as three months. But in the first two years of Zuri's life, I'd only been on a couple of expeditions, and both were much shorter compared to what I'd done before she was born. Although both trips were wonderful in their way, they were accompanied by the emotional guilt of being so far away from my daughter.

The reality of my life at that time was that I worked full-time and Oba worked even more, from 9:00 a.m. until 9:00 p.m. most days. Since I didn't make much money in my position at the museum, we prioritized his work over mine because his job paid the bills and kept the roof over our heads.

Though it's somewhat embarrassing to admit, at times I used to daydream about not being married and not being a mother. I felt selfish for continuing to want to have this adventurous wildlife career while also being responsible for a child.

One day around this time, I was having a difficult conversation with Logan. I was subtly trying to navigate our relationship away from the physical and back to the professional, in light of my suspicions that he wasn't who he seemed to be, and he was simply trying to figure out how to see me again. He said he knew people who were studying jaguars in Panama and suggested that might be a great place for me to get some experience with different carnivores.

To me, jaguars are one of the most fascinating big cats in the Americas. More important, jaguars are an endangered species. So the idea of actively working on the science that could help bring their populations back was appealing.

The best part was that all the details were already worked out. Logan said he had it all handled—he would arrange everything, which was a huge relief for me and made it much easier to say a resounding yes to the expedition. Additionally, since the expedition would include several other colleagues, it appeared as though this trip would have to be strictly professional, which I realized was beginning to feel like the right thing again. I found myself believing that I could erase the betrayal I'd allowed myself to be a part of for the majority of 2017 by going back to doing real science work with Logan and groups of scientists, leaving the illicit romance behind. While Logan worked out the details, I was able to focus on what to do with Zuri for the two weeks I'd be gone.

In the past, I'd heavily relied on my mom. So I called her and told her about this awesome opportunity in Panama. She was excited for me, but when I gave her the dates, she said she had a big commitment she couldn't work around and therefore couldn't help very much with Zuri.

"Oh, don't worry about it. I'll discuss it with Oba, and we'll figure something out," I reassured her.

When I did ask Oba to help me figure out a way to provide childcare for Zuri while I was in Panama, he rolled his eyes.

"No, Rae," he told me. And in that no, he not only declined to help me find a childcare solution but also refused to say yes to the whole expedition. He wasn't excited about the trip. He wasn't excited about me having a potential once-in-a-lifetime chance to study endangered jaguars. He said no—end of discussion.

I'd expected him to be unsupportive, but hearing his flat-out refusal still hurt. And it made me wonder if he was picking up on the fact that Logan was more than a platonic colleague.

For whatever reason, I was unwilling to take no for an answer. I refused to let parenting be a barrier to my career success and my joy. Somewhere inside of me, I thought that if I didn't go on this Panama trip, if I allowed myself to miss this opportunity, it would cause a domino effect that knocked down the rest of my hopes and dreams. Maybe this would be the first of many noes that would eventually obscure the path I wanted to be on.

I'm not proud of my thought process during this time. Although I applaud myself for being determined to figure it out, I also wish I could have given myself a moment of stillness to really listen to my heart and understand what it was after. Yes, I wanted to stretch and grow and have new experiences. However, doing any of that with Logan not only was distasteful but also was beginning to cause me to hate myself.

Then an idea occurred to me: *What if I bring Zuri to the field with me?*

I had never heard of a scientist who'd done fieldwork in any capacity and brought a toddler along. So I don't know what led me to think this could be even a remotely good idea. Yet I believed it could be an opportunity for me to do something new, something I'd never seen modeled by anyone else. It also could be my ticket to freedom, the thing that changed Oba's mind and demonstrated that I was capable of succeeding as a wife, a mother, and a badass field researcher, all at the same time.

When I broached the subject with Oba, he said that even though he didn't like it, I could do it. He seemed resigned and disinterested in how it would go, and I felt like this was my opportunity to both prove to myself that I could act appropriately around Logan and handle the responsibilities of parenting and science exploration simultaneously.

I took a picture at the airport, with Zuri in her stroller and me holding her favorite stuffy, a giant orange Popsicle that we called Creamsicle. I'm wearing a backpack and have a diaper bag slung across one of my shoulders. I posted the photo on social media with a caption that said something along the lines of, "When you don't have a babysitter, but you have to go to the field to study jaguars."

We stepped off the plane in Panama and exited the airport into the oppressive, humid tropical air. Logan was there to pick us up, and after platonic hugs hello, we piled into the car. Several hours later, we arrived at our destination, a village at the edge of the Gamboa Rainforest.

Locating jaguars is unbelievably difficult. First of all, the jungle is super dense, which makes it hard to find anything. And second, jaguars are elusive. That's what makes them excellent predators—they can sneak up on their prey without ever being seen. Jaguars also are territorial; there's only one jaguar per region.

Our plan was to look for evidence of jaguars' presence, like poop or a recent kill. Then we'd set some camera traps, capture images, trap them, and put GPS collars on as many of them as possible, so we could understand their movements and habitat preferences. Knowing the areas they used could help us recommend areas for conservation.

The next day, I woke up early, grabbed Zuri, and took her into the village. We needed to meet with potential caregivers to keep her safe while I was in the rainforest. (Toddlers and big carnivores don't exactly mix well.) It didn't take long for me to interview someone I felt was perfect for the job, and I hired her on the spot. With that squared away, I was ready to switch my focus to finding jaguars. We needed to meet with the researchers who had already mapped out the jaguar study.

"Okay, what's the plan? When are we getting out there?" I asked Logan as we sat down for our inaugural planning session.

It turned out that he hadn't heard back from any of the researchers yet, which was highly abnormal. You don't set out on an expedition without communication from the research team. But I still naively trusted Logan, so I figured I needed to be patient.

Over the next couple of days, it was delay after delay after delay. All he had to offer me were excuses: "Oh, they spend so much time out in the field. It's difficult to reach them when they're not plugged in. They're always really bad at getting back to me about this stuff."

These delays swirled up tumultuous emotions within me. I was angry, terrified, and suspicious. Had I dragged my two-year-old daughter and myself all the way down to Panama for nothing?

But at the same time, my optimism and joie de vivre kicked in. *You know, we're already here*, I thought. *I might as well make the best of it.* We were still living in a house in a village on the outskirts of a rainforest. Just taking a walk down the street with Zuri was a magnificent adventure of unique sights, sounds, and smells.

One realization I had during these days of waiting was that no one in my family, including Oba, had ever come with me to the field. It meant so much to me that, of all people, my daughter was the first one to go on such a trip with me. She was able to witness firsthand what Mommy was doing when she traveled for her job exploring wildlife. We were experiencing and learning about a completely new location for both of us, together.

It was a parenting highlight, to be able to show Zuri something that I didn't experience while growing up: that she belongs in all kinds of spaces, including the outdoors. And I hope that, even though she was so young and doesn't really remember the trip, that experience translates into her growing into a woman who feels strong in making her own choices. That she won't be afraid to do something brand-new or nontraditional. That she recognizes it might be scary and she might feel isolated sometimes, but she's capable and can also bet on herself.

About a week later, Logan still hadn't heard from the researchers, and we hadn't gone into the field. I'd spent hundreds of dollars on childcare for Zuri and spent loads of time taking hikes through the rainforest and getting to know researchers on other projects in the area, but nothing was happening with the jaguar project. One day, I put Zuri down for her afternoon nap and went to find Logan. It was time for a serious confrontation.

"I did not fly all the way from New York to Panama with my toddler, putting my whole marriage in jeopardy, to come back empty-handed from this whole thing," I told him. We had to make something happen, and we needed to get our butts into that rainforest.

"Okay, you're right. Let's at least go and try to find these researchers," he replied, as if this idea was only now dawning on him. I wasn't satisfied by this new plan, and as each day passed I was feeling more and more trapped and deceived. I wanted the opportunity to be real, and I wanted to walk away from this with success, so I agreed to go on a search for these researchers. The next day, we set out into the rainforest.

If someone asked me what the rainforest sounds like, I would say that it's noisy, but in a consistent way—the squawks of birds, the hum of insects, and the sound of raindrops falling. As we passed through, the howler monkeys made their displeasure known. I was under their tree, and they didn't like it one bit. They threw fruit and other objects down from high up in the trees, screaming at the tops of their lungs. Then the whole group would run away in a loud frenzy.

For hours, Logan and I hiked through the muddy rainforest. Even though we were alone, we weren't romantic—a distrust was building within me, and it was driving a wedge between us. We saw neither the researchers nor a single human being. We didn't even see evidence of these people: no camera traps, no gear, no flagging on the trees.

"Listen, we're never going to find these guys. We have to get back home before bedtime. This whole mission is essentially a huge failure," I said, accepting defeat.

We made our way down along a path, which eventually ended. Then we reached a shallow stream where we could wade across. So they'd stay dry, I took off my shoes and rolled my pants just above my knees.

The banks of the stream were covered with mud. Right before I put my foot down in the mud, I saw what was unequivocally a jaguar print.

It was a classic big-cat paw print, a much, much larger version of a house-cat print. I could even see the claw markings. As I stared at the beautiful, rare sight, I realized it was evidence that a jaguar had been there recently. It had rained only thirty minutes before my discovery, so any prints were guaranteed to be extremely fresh.

I instinctually pulled out my GPS unit and took a point, to record the longitude and latitude of exactly where we had found the jaguar print. Of course, we had nothing to do with that information, since we'd never connected with the researchers. It was frustrating, but I was still glad to have a fun data point to take home with me.

In that moment, I was in the closest proximity I had ever been, and maybe ever would be, to a jaguar in the wild. A jaguar that, for all I knew, could have been watching me the whole time. Even though I never laid eyes on the animal itself, I felt that the print was its way of telling me it was there, elusive as ever, and that it and its habitat needed to be protected.

When we returned from the field that evening, I changed my plane ticket so that Zuri and I could leave the next day.

I told Logan, "Do whatever you want. Stay here. Don't stay here. I don't care. But Zuri and I are out of here. I appreciate what you tried to do, but this isn't working out. This isn't what we agreed on." I felt that I needed to step up, be a responsible adult, and get us out of there and back home safely. I was angry and firm with my words and felt proud of myself for course-correcting a trip that perhaps should never have happened. It still wasn't obvious to me whether Logan had lied about the jaguar-research opportunity as a way to manipulate me into going on an exciting trip with him or if he wasn't to blame for things falling through at all. But in the end it didn't matter. I was making decisions for more than myself—I was making changes to protect Zuri, to protect my time, and to protect what was left of my own dignity. I wanted out of the affair with Logan, I wanted out of the trade-off between exploration and parenting, and most of all, I wanted out of Panama.

On the return flight, Zuri fell asleep, using Creamsicle as a pillow. As I gazed at her sweet, innocent face, I acknowledged that no matter what—even when I was having adventures without her—the two of us were inseparable. I clung to the confidence that Zuri and I would be okay, as long as we were together.

Life in the wild was often unpredictable. But I knew with certainty that I could take care of myself and my daughter. That even if I couldn't depend on anyone else, I could depend on myself.

*   *   *

DURING MY TIME IN PANAMA, away from the rigors of daily life, and without the intimacy Logan and I typically shared, I had a lot of time to think. One of the things I thought about a lot was how to break it off with Logan. The honeymoon period of our entanglement was waning. I was now more able to see that a relationship with him wasn't better than my marriage—he mostly answered my need for sex.

I could feel myself screaming internally, telling me to turn around and go back toward good decision-making. I felt like the people watching a horror movie in a theater and yelling at the screen: *"Don't go in there!"*

Fear slowly consumed me. I had crossed a line that you can't uncross. I'd given up my body, and my mind was telling me that I'd also given up my autonomy, my freedom, and my self-worth.

Logan must have sensed that something within me had shifted because he started to pressure me into leaving Oba. He suggested that we could both get divorced and ride off into the sunset together.

Questions swirled in my mind. By betraying my husband, didn't that mean I was trapped on a path toward a life with this other man? That the man you slept with outside of your marriage was the person you were destined to be with the whole time? That throwing your former life in the dumpster freed you to finally have the happy ending you deserved?

If that were the case, why did I feel like I was locked into a roller-coaster car, hurtling down a track that ended at a brick wall? I dreaded that painful crash at the end.

I also was in denial. I believed that I had truly never gotten it right—I had always settled, always people-pleased, always found my own value in how much attention a man was giving me. And on top of the shame of that realization, along with the shame of committing what our society deems a major sin and dragging my daughter into this toxic ocean of chaos . . . I was repulsed by who I'd become, yet scared to revert to who I'd been before. I was a body of lies and, worst of all, I was lying to myself.

I needed to clarify what I wanted and ask myself questions that would lead me in that direction. Did I want to have a shell of a marriage with Oba and get validation from elsewhere? No. Did I want to get divorced and be single? No. Did I want to be in an official relationship with Logan? No. So what *did* I want?

THINGS CAME TO A HEAD in October 2017, when I was at an extremely low point mentally and emotionally. I found an excuse to go to the Adirondacks in upstate New York, to meet with a conservation group about a black bear project while also secretly spending the weekend in a cute Airbnb with Logan to celebrate my birthday. The trip happened to overlap with my period. Typically, my periods are heavy and oftentimes painful. On my heaviest days, I like to rest as much as possible—and I'm certainly not interested in sex.

When we arrived at the Airbnb, Logan tried to make a move on me. I reminded him that I needed to rest because my period was draining me. He was disappointed and pressured me so heavily into having sex that for the first time, I clearly saw that whatever he was giving me, it wasn't love. Our relationship was transactional: he validated my dreams, I gave him my body. I should have packed, gotten in the car, and driven back to New York City, but I didn't. I stayed the full weekend, weakly voicing that I didn't want to have sex and feeling guilty when he was angry about it, over and over.

By the end of the weekend, I felt utterly disgusted, with both him and myself. Logan had gone from worshipping me in the forests of Madagascar to disrespecting me deeply in less than a year. And by the end of the year, in spite of Logan's behavior, I couldn't deny that I felt more alive in his entanglements than I ever did in my crowded life in New York City. In a matter of months, I had gone from a woman searching for purpose to a woman who'd lost all dignity and self-respect. I was in crisis, and nobody knew.

One afternoon, as I was working at the museum, I received an email from an old acquaintance: Leona.

Our relationship hadn't been friendly, and my mind flashed back to that last night in Madagascar, when her jealous gaze had burned through Logan and me. Reluctantly, I opened the message. Scrolling past the obligatory salutations and pleasantries, my eyes were alerted to a mention of Logan. Many mentions of Logan.

Leona provided previously undisclosed context to our time in Madagascar. She detailed the first time she'd met Logan on the island—months before the rest of us arrived, when he was there scoping out the forest and making preparations for the future expedition. Logan and Leona started hooking up, and he led her to believe they were a couple. She was a young woman watching in anger and confusion as Logan turned his attention to another. What I had perceived as jealousy was anger, hurt, and fury at the man she had once mistakenly loved.

Logan had blocked all contact with her, but she was desperate to reach him. So she'd contacted me—seeking solutions to a problem I couldn't solve.

The world around me stopped. The life I'd been living for the past year was a huge, incalculable lie. My desire for Logan's validation and his belief in my pursuit of success had blinded me to his true nature. In the wreckage of our affair lay three broken lives. And a dream that once felt so close felt further away than ever.

I replied to Leona in a brief message: *I am so sorry I didn't realize what was happening when I was there. Thank you for contacting me. I hope this can all be made right.*

Next, I called Logan and told him we were done. He threatened to kill himself. I told him I couldn't save him. I deleted his number and prepared for a hard conversation with my husband.

Walking into the apartment, I found Zuri asleep and Oba reclining casually on the couch. As he sifted through emails, he greeted me with our usual, lackluster "welcome home."

I sat beside Oba. On the train ride home, I'd written and rehearsed my self-composed script innumerable times. But when I opened my mouth to speak my lines, I began to cry uncontrollably.

"I am a piece of shit," I said between gulps of air, and then I managed to string together quasi-coherent sentences rehashing the past year's events.

Throughout my tortured confession, Oba remained quiet and stoic. When he finally spoke, it was a haphazard barrage of questions and accusations.

Much was revealed that night, and little achieved. On the surface, we were the epitome of Black excellence. Yet behind the scenes, turmoil lay waste to the shreds of ascension to which we clung.

We decided to separate, not knowing what came next. We sat in silence, the rubble of our selfish, reckless decisions strewn around us.

*   *   *

OVER THE NEXT SEVERAL MONTHS, Oba and I somehow continued to cohabitate and co-parent; for the time being, we'd decided not to separate after all. In the close quarters of our Harlem apartment, we communicated, rekindled, and reconnected. We also dedicated ourselves to marriage counseling. We promised each other that although we were in a breakup posture, we owed it to ourselves, and to Zuri, to fight for our relationship. We agreed that we needed to at least work hard at fixing things before we let it all go. That took a lot of maturity and strength, and I'm so impressed with us that we did it. A lot of truth emerged from those months. Honesty became our guidepost, as I unpacked what had led me to Logan and what I needed from Oba.

In those sessions, I learned that it wasn't that Oba was completely disinterested in my work—it was more that he didn't agree with the fieldwork aspect of it. At one point he said, "I don't want a wife who travels! I want a wife who works in an office, in New York, and comes home after work every day." He was a professional and wanted a counterpart with a complementary career. The image of the Black power couple was front and center in his mind. And with that image came ideas of a wife who could also be consistent in the home, creating what he saw as a strong foundation for raising children—a role I appreciated but which wasn't the right fit for me.

Counseling provided Oba with space to be completely forthright, and he was able to voice some important truths. Sometimes it was painful to hear these realities from my husband, but the relief of finally having clarity was tremendous. It helped me see that I myself wasn't the problem. The way he treated me wasn't personal. He straight up wanted a whole different kind of wife. I realized we weren't compatible. While the cheating was my fault and our breakup was my fault, the impact our incompatibility was having on our marriage was due to our lack of voicing what we each wanted and needed from the relationship. We should've been having these conversations long before he ever put a ring on it.

I admitted that what I wanted and needed was to be out of a relationship for a while, to regroup and eventually—hopefully—find someone who was excited about the nontraditional nature of my professional passions. I also needed someone who was interested in traveling with me, who would encourage my pursuits and partner with me on going even further down all of my paths. Someone like Logan.

Despite everything, I missed him.

While he was monstrous, no one had believed in me like he did. No one listened to me like he did. And no one encouraged me like he did. I didn't even have that same confidence in my own abilities—a truth that stung to admit.

Prior to my Madagascar trip, I felt lost in the world I had created for myself. The prescriptions of the everyday became rote and uninspiring, and the antidote was Logan—less so the man and more so what he represented and what he inspired in me. Even after all I'd endured because of the affair, I clung to him and the memory of who I'd been when we were together like a fading dream.

WHILE I SIFTED through the ashes of a home life I'd taken a match to, I leaned in to my work at the American Museum of Natural History. However, as 2017 flowed into 2018, it seemed that conflict was determined to erupt in every sphere of my existence.

"Scientists only. Ma'am? This floor is science-staff only," said the woman blocking my exit from the elevator by flinging her arms wide and using her body as a barricade. When I kept trying to walk out of the elevator, she pressed her hand on my chest.

"This floor is not open to the public. This is a floor where scientists work—you need to go back down."

"*I am a scientist!*" I basically screamed while pushing past the woman and bumping her into the side of the elevator. I rushed forcefully into the hallway of the fifth floor.

"*Excuse me!*" she shouted in reply. Her yell followed my hurried steps as my students obediently followed me, and once we had rounded a corner, I stopped and apologized to them. They were Black tenth and eleventh graders from the local public school, participating in an internship program where kids spent a couple of afternoons a week helping a museum scientist with their research. I'd specifically asked for Black students to mentor and had been working with this group for several months. Although I'd experienced all kinds of mistreatment at AMNH over the years, this was the first time my students had witnessed me facing racism at my workplace.

I had been a scientist at AMNH for almost three full years. My title was "Conservation Science Research and Teaching Postdoctoral Fellow," but my job was to conduct research on bears and teach some advanced classes for the museum's educational programs. My office was on the fifth floor, where all the other science offices were, and my colleagues were world-renowned scientists. And I was among them. We all had our PhDs and were engaged in high-level research. But I was the only Black full-time biologist. The same time that I was at AMNH, Neil deGrasse Tyson was the head of the Hayden Planetarium, housed in the same building. When I met with him once for general career advice, the first thing he said was, "I'm so glad you're here! Now I'm not the only Black scientist anymore."

My commitment to pursuing a career combining science and communications led me to seek a publicist. I wanted to work with Gilda Squire, who was well-known for leading the publicity efforts of superstar Misty Copeland. I had pitched myself to Gilda as "the Misty Copeland of wildlife," which caught her attention. Gilda, understanding my limited finances, offered me a deal: for a reasonable fee, she'd secure two television interviews and two printed news interviews for me. And she delivered. That week, I was a guest on WNET's *MetroFocus*, a local morning news show that interviewed me about the uniqueness of my work: both my bear research and my nontraditional identity as a Black woman in wildlife science.

From that point onward, anytime I was interviewed, the journalist would note that it was uncommon for a millennial Black woman—or any Black person at all—to be a wildlife ecologist and study bears. I'd say yes, I didn't know of a single other Black woman who did the same kind of work that I did. And I'd add that at AMNH, I was the only Black full-time biologist.

Before long, word about my media appearances got back to the AMNH communications department, who notified my boss about what I'd been saying. Shortly thereafter, I was called into a meeting, and my boss confronted me about it.

"But I am the only Black biologist here," I said to my boss.

"I know, but when you say it, it makes the museum look bad," my boss asserted.

"Well then, y'all should hire more Black folks in science positions."

"It's not that easy."

"What do you want me to say then? Do you want me to lie?"
I wondered aloud.

The question hung in the air between us, the tension thicker
than ever.

"I'm not saying you should lie," my boss finally responded.
"But you need to find a more careful way of communicating
with the press."

That wasn't the last meeting we'd have about the matter. I
was angry because I believed I was being censored specifically
because I was a Black woman wielding her power and growing
in strength. To me, it felt as though my boss cared about the
reputation of the American Museum of Natural History and her
job security more than protecting me from discriminatory treat-
ment. In both my personal and professional life, I was no longer
willing to suppress my truth to allow everyone around me to stay
in the safe cocoon of their comfort zones. I understood that path
forward had consequences, but as my disintegrated marriage and
increasingly toxic workplace showed me, the cost of my silence
was far too high.

Around the time I decided to leave my AMNH job, I wrote in
my journal:

> I care more about black and brown liberation than I do about
> oppressive institutional hierarchies, and it's causing me to have
> to choose between a safe job doing what I love for a prestigious
> institution and my own dignity. It took me a few years in the
> system to realize it's easier for someone like me to forge a path
> in a dense rainforest than it is in a bureaucratic institution,
> even if the goals—saving species from extinction—are the
> same in both places.

*   *   *

IN ADDITION TO marriage counseling, I found an amazing therapist whose office was only a few blocks from the museum. Once
a week during my lunch break, I'd walk over to her office for a
session. She was a beautiful Middle Eastern woman who was less
than ten years older than I was. She usually didn't say much during
our sessions, but she'd ask general guiding questions and had an
almost magical way of drawing profound conclusions out of me.

As disappointed as I was in myself for contributing to such
a mess and causing so much heartbreak, a tiny, forgiving voice
in my head would speak from time to time and give me grace.
Perhaps my connection to wild animals was showing up in this
stressful situation. In the wilderness, when it comes to fight or
flight, one is not superior to the other. One is not more noble,
and one does not result in greater success. Fleeing the scene is a
highly respectable, and often more effective, survival strategy.

I wasn't absolving myself of wrongdoing, because the mistakes
I made were glaringly obvious and causing my whole world to
crash down. Yet I was able to see and have compassion for the
young woman who had made those mistakes. A young woman
who had spent years trying to convince her husband, and herself,
that she'd dim her light to make everyone else feel more comfortable. A young woman who was too afraid to trust herself and
continue down such a nontraditional track.

For so long I had been blaming Oba, telling myself and telling
Logan that I'd been trapped in my marriage. But while working
with my therapist, who often sat in silence as I spilled my soul
onto the office floor, I finally said the truest words I'd ever spoken: "I trapped myself."

It was immature to blame Oba for the unhappiness I'd been
experiencing in our relationship. He was at fault for many things,

but throughout our dating, engagement, and marriage, he was consistent. He was the same man, through and through, from the day I'd met him in November 2010 to the day I told him about my affair in December 2017. He'd always been up-front about who he was, what he wanted, what he was comfortable with, and what his goals were.

I'd been overlooked by so many smart, attractive, eligible Black men for years, so Oba's early interest in me clouded my vision and my sensibility. Although I might have convinced myself I could make it work, I didn't want to be in a relationship that valued tradition and comfort zones over exploration and adventure. I wanted a supportive partner who loved my freedom as much as I did. Who saw all that was nontraditional in me as qualities to love, not things to settle for or try to dismantle. And more than anything, I wanted to raise children with someone who was excited to see how we could make it all work. I also made peace with the fact that I wanted a lot of attention from my partner. For years I'd perceived this as a flaw, some kind of needy immaturity. My therapist helped me realize this part of me wasn't a defect.

I had to take responsibility for the fact that, in many ways, I had ruined my own life—my marriage, my role as a parent, my professional reputation, my stable lifestyle, the respect of my family and friends, and beyond. That was scary and depressing. I regretted my decisions, hating myself, and was overcome by shame. I couldn't eat, couldn't focus at work, couldn't look my parents in the eye.

As challenging as accepting these truths was, one of the biggest struggles I faced was that for the first time in my entire adult life, I couldn't go to nature to heal. After being the villain in my own life story and hurting so many people so deeply, I couldn't call in another favor, ask Oba or my mom to look after Zuri, and travel to the wilderness, even though the wild was the place that had always offered me healing and sanctuary.

After months of working with my therapist, I finally felt I was healing my relationship with myself and rebuilding my self-esteem. In one session, I mentioned to her that Oba and I might be separating, as our marriage seemed unsalvageable. Yet I had doubts about whether that was the right move. She fidgeted in her seat as if she were uncomfortable, and by the end of our session, she blurted, "Rae, you've made so much progress. Don't go backward now—you know what you need to do."

I was grateful for her words. As ambiguous as they were, I like to believe that we both knew what decision I needed to make.

IN THE SUMMER OF 2018, I resolved to quit my AMNH job and move to Washington, DC. For years I'd felt my best employment prospects were in DC, because that's where most environmental NGOs (nongovernmental organizations) were headquartered—and I had my eye on National Geographic in particular. Regardless of whether my relationship with Oba healed, I needed to work toward my professional goals. I felt that doing so could be part of the answer to finally feeling like I was living life for me.

Not surprisingly, Oba didn't want me to leave New York, and he certainly didn't want me to relocate with Zuri. It took a lot of discussion and cooperation to reach an agreement.

We decided to separate first, but then we began the divorce process, all the more painful with Zuri at its center. I was most concerned about her maintaining a healthy relationship with her dad. Oba was and still is a good father to her, and I felt guilty that she was losing the ability to see her father every day and to grow up with her two parents together. But the part of me that was healing would always chime in and remind me that Oba had the ability to be an excellent father to her under many conditions. Not only this, but a mother who was happy, healthy, and well-resourced was one of the best things Zuri could have, and it was something that would provide significant, lasting positive effects.

The broken part of me remained lost and alone and sad, far from the reflective respite of the wilderness, laid bare in Upper Manhattan's asphalt streets. I had spent years feeling trapped. But what I came to realize was that I was trapped by the idea that life comes with rules. I was a part of an educated, middle-class Black community of traditional professionals—lawyers, businesspeople, doctors—all Oba's friends. But why did I refuse to see all of the diverse lifestyles around me?

I didn't have any examples of Black people, let alone Black women, having professional expertise in the remotest, wildest destinations. I didn't have any examples of the scientists I was surrounded with leveraging media to connect with larger audiences. I didn't have any examples of mothers of young children who left home for months at a time to look for endangered species. But I had a semblance of some of those things. And I had myself. And that's who I was refusing to see more than anyone else: *myself.* And if I couldn't see myself, cherish myself, or believe in myself, then how on earth could I expect anyone else, even my husband, to do so?

The pain lingers, for me and for Oba, and co-parenting remains a roller coaster that luckily has more highs than lows. But I began living, and will continue living, in the light—without secrets or lies, to my partner or to myself.

Chapter 11

I'd left Washington, DC, in the summer of 2008 to earn my credentials and become the scientist I dreamed of being. And here I was exactly one decade later, moving back to the nation's capital, three-year-old daughter in tow, with science degrees, world travel, love, and heartbreak all packed in my metaphorical baggage.

H Street, formerly a Black commercial neighborhood in northeast DC, has since become a neighborhood of towering luxury apartments, rock-and-roll-themed bars, and high-end yoga studios. While some Black homeowners and businesses have managed to hang on to the neighborhood's nostalgic aspirations, many of the last Black-owned businesses closed shop during my tenure there, citing out-of-control rent and inflation. Zuri and I, having just moved to the neighborhood, lived in a new, fancy apartment building with a concierge, rooftop swimming pool, and majority-white neighbors. My daughter and I were a strange mix of so many demographics—privileged in some ways, underprivileged in others; educated yet low-income; professionally experienced yet entering unfamiliar territory—and it felt like we reflected the block's changing, dynamic landscape.

I'd been saving money, which was a feat considering my research fellowship at AMNH had paid sixty thousand dollars a year for the past several years. That money might go far in some places, but not in New York City. After taxes were deducted, as well as the health care benefits we used for the whole family, my paycheck barely covered the expense of full-time childcare and my contribution to rent on the apartment Oba and I shared. I'd been intentional about saving every dollar I earned from other sources—public speaking engagements, adjunct-faculty teaching positions, and minor media opportunities. Those savings were coming in handy now as I forged my path in Washington, DC.

Although DC was full of environmental nonprofit organizations, even those I'd worked with before, I had one objective: work with National Geographic. Logan had put this idea into my head. He'd been awarded research grants from National Geographic several times, and I'd seen firsthand how much that affiliation had benefited him. I loved all the ways that the group offered the perfect marriage between science, media, education, and outreach, the mix I'd been seeking for years. However, instead of taking matters into my own hands, I'd foolishly waited on Logan to make the introductions that would get my foot in the door.

In the weeks leading up to my move to DC, I marched myself to the Nat Geo headquarters and convinced a grant manager to meet with me. Having emailed him over and over again, I'd beaten his patience to a pulp. Sitting across from him, I used photos and stories from my years in the field to evoke an imaginative future of conservation and ecological research that, with the help of National Geographic, I could achieve.

And it worked.

So began my role as a verified National Geographic Explorer. With the help of a research team and grant money, I designed a project fellowship with a strong reliance on media. I negotiated a salary that allowed me to survive as a single mother in DC and orchestrated a job that allowed me to travel the world.

Although I had enough money to pay the monthly rent of three thousand dollars, I didn't have enough to pay for full-time childcare and pay down my constantly increasing credit card bills. I was both an upwardly mobile Black professional and a single mother who would forgo necessities for myself in order to feed and clothe my child.

I was straddling two different worlds. On one hand, I was an Ivy League–educated, trained biologist in her early thirties who had caught the attention of National Geographic. On the other hand, I was a single Black mother who was overworked and underpaid, trying to establish a nontraditional career in a system with unequal barriers to advancement. Even so, that luxury apartment I miraculously was able to lease nurtured Zuri and me for two beautiful years, while I shored up my self-esteem, independence, and budding science-communication career.

A couple of weeks into my new DC life, I was tasked with my research assignment from National Geographic, Zuri was in preschool, and Oba and I were communicating better than ever. Despite so many things falling into place, anxiety and depression paralyzed me. For several days in a row, I walked Zuri to her preschool three blocks from our apartment, kissed her goodbye at the classroom door, walked back to my apartment, crawled into bed, and lay there, in a daze, until I had to pick her up at 3:00 p.m. I didn't feel good; I didn't feel bad. I didn't feel at all. My mind was numb, and I had no motivation.

I'd been hustling to land on my feet after leaving everything familiar back in New York, and in the months since I'd admitted to my affair, I'd kept myself busy with a life largely away from home so I wouldn't have to sit still. Back in New York, I woke up early to work out three days a week, pulled long hours at the museum, and traveled with Zuri to visit family in New Jersey or

California most weekends. Now, I was alone. No office to rush to, no husband in the next room to avoid.

Scoring a fellowship with National Geographic was meaningless if I didn't prove myself to be outstanding, and that meant hitting the ground running and producing amazing results with my research. Yet I couldn't even get out of bed. I was depressed, ashamed that I was depressed, and withdrawn from friends and family, even contemplating whether I could fall asleep and never wake back up.

In that scary moment, I thought of Zuri. As much as I no longer wanted to feel pain and pressure, the last thing I wanted her to see was a mother who gave up.

I needed to work on myself. And not solely for Rae, but for the other people who relied on me. If I was going to make it, I had to understand the root of my pain, take time to process the traumas I'd endured, properly accept my new reality, and build the strength I needed to thrive at this point in my life.

Into therapies I dove: the classic kind, the alternative kind, the book kind, the podcast kind, the Oprah kind. In every spare moment of the day, I listened to people testify to the rough times they had endured and the tools they used to overcome.

I also found the strength to seek out a new therapist. The sessions were one hundred dollars each, and I had to pay for them out of pocket. So I was only able to go twice a month. The expense was far outside my extremely tight budget, but I made it a priority. If I had to max out my credit cards to keep myself alive, then that's what I would do.

My DC therapist was an incredible Black woman about my age. She helped me explore the questions that were blocking my ability to thrive from the inside out: why I projected strength but

was filled with fear, why for so long I had sought external valida-
tion from men, and why I was obsessed with my professional life.
I didn't know anyone as obsessed with their career as I was, and
I struggled to separate my worth from my work.

I challenged myself to recenter joy in my work. And in doing
so, I found my purpose and my people.

My career blossomed in ways I had previously thought unimaginable. For the first time, I was dictating where I traveled and
what I studied. With each new expedition, my belief in myself
grew, and I, in turn, started to heal.

Simultaneously, as Oba and I solidified into separated
co-parents, a new relationship blossomed. Dave, my dear
friend and former colleague at the American Natural History
Museum, was slowly coming into my focus as more than a
friend. Tall, strong, strikingly handsome, creative, and decidedly feminist, Dave was very desirable, and even more so because
we had built a strong and trustworthy platonic relationship over
the several years we'd worked together at the museum. He had
been the first friend I'd told about my affair and the shame I
carried, to which he responded with compassion, empathy, and
understanding. Although we had never crossed a line when I was
living in New York, it was during my first months in DC that we
acknowledged the way we missed each other perhaps signaled
that we had something deeper than a normal friendship.

Mutually respecting how fresh my separation from Oba was
and the healing I still needed to do, Dave and I took it slow at
first. Many nights we'd check in with a brief FaceTime chat,
rather than the hours on the phone that I used to associate with
a budding relationship. When I felt ready, he started visiting
me in DC on weekends when Zuri was with Oba. Dave and I

would take short road trips to forests and lakes in Maryland and Virginia, spend the night at an Airbnb, and return the next day. Dave would then hop on a train back to New York.

From the beginning, he was unlike anyone I'd ever been with. Dave was romantic, and not in a goal-oriented way. He treated me like I deserved to be cared for—something I believed but had never actually experienced. Whether with a foot rub, a home-cooked meal, or a song, Dave demonstrated his love to me without expecting something in return.

Having been previously married himself, Dave understood the journey I was on, and he made space for me to find my footing without exerting any pressure. He was already sexually irresistible, and his respectful, considerate, mature, and vulnerable nature made him even more so.

AT THE BEGINNING OF 2019, I was several months into my National Geographic fellowship studying grizzly bears in Montana. I was also working as an adjunct professor at Johns Hopkins University, teaching principles of ecology to graduate students.

Of course, being an adjunct faculty member isn't a secure position. To create more stability in my career and to demonstrate my value to the department and the institution, I decided to pitch a graduate-level field course in wildlife ecology. If the field course became popular, then I'd be on more solid ground.

After a successful pitch to the department head, I was left to plan the details of the course. I decided not to limit myself to places and species that were familiar to me. Where had I always dreamed of traveling? What animal had I longed to study in the wild?

Lowland gorillas in Central Africa. That's what I wanted to build my proposal around.

The idea presented a double opportunity. I'd launch the field course for the students at Johns Hopkins, and then through the course I'd have access to the site to start a gorilla research project. Gorillas are highly endangered in Central Africa, and their habitats are also under threat. If I could establish a research project there, it would be one of the first studies in that area, and we could learn a lot about where the gorilla population lived and what needed to be done to protect them, so they could grow and thrive.

I invested a lot of time in putting together a proposal and a syllabus. Soon I had to present my idea to both the department chair and the dean of the school. When I shared my vision of the project, they loved it. This was something vastly different from any other field course they currently offered.

On top of approving the proposal, they gave me funding to do a two-week reconnaissance trip to scout the field site in a rainforest in the Congo Basin called the Dja Faunal Reserve. This would allow me to ensure that the location was optimal for both the field course and my research project.

From a career perspective, this was incredible. But I underestimated the toll this potential adventure could have on my personal life and on my mental health.

This trip would be my first long international expedition since my separation from Oba. I couldn't up and leave. I'd be in an extremely remote place with zero contact—no electricity, no Wi-Fi, no cell phone reception. I'd be totally cut off from my daughter and her caregivers the entire time.

In that moment, a brand-new thought occurred to me: *It is too much.* Never in my life had I considered that anything would be too much for me. If anything, I'd always felt that I could be doing more. I had told myself that I was better off as a single woman, mothering on my own, and that I really could do it all.

Now, the moment had come for me not to just talk the talk but also to walk the walk. And I absolutely freaked out.

To combat the encroaching feelings of overwhelm, I knew there was only one thing I could do: overprepare. I needed to spend as much time and energy as possible planning for my two-week expedition.

So that's what I did. I made sure everything was taken care of for the trip, from all the transportation to how many pounds of rice I needed to carry in my backpack to survive at the field site.

And even more than all the expedition logistics, I nailed down exactly what was needed for Zuri to be safe and nurtured while I was on the other side of the world. I had a whole plan and system in place for her to be cared for, including being dropped off at and picked up from preschool every day.

One of the most critical pieces of preparation for the expedition was finding the gorillas. I'd done some primate work before, so I fully expected that it would be as challenging to find these gorillas as it had been to track down the lemurs.

After I did a ton of research, I learned about a group of people who lived right outside this rainforest, called the Baka. If anyone was an expert on gorillas in that area, it would be them. I tracked some folks down and asked if I could hire someone as a guide on this project. A couple of men agreed, and they assured me gorillas lived in the area and they could take me to them.

I was thrilled that everything was falling into place.

But then, a couple of weeks before I was scheduled to leave on the expedition, I got word from the staff at Johns Hopkins about a rule that would necessitate a major change to my plan: any travel done under Johns Hopkins University's name required the participation of a full-time faculty member.

Thankfully, the department chair found a full-time faculty member willing to accompany me. Her name was Jen, and my first impression of her was that she was prim, proper, and petite— definitely not someone I'd expect to volunteer to go on a Congo Basin expedition. Funny, considering how often I myself hadn't been what people expected throughout my entire career. I felt that she wasn't the right person for this expedition, but what choice did I have? Too much was on the line, and I couldn't find another volunteer in such a short amount of time. So a few days later, Jen and I flew to Yaoundé, the capital of Cameroon.

The Congo Basin is a humongous region that encompasses a large portion of Central Africa and spans across multiple countries. Our expedition would take us to the section located in Cameroon, to the Dja Faunal Reserve.

When Jen and I arrived in Yaoundé for the recon mission, the thick jungle enveloping the airport hearkened back to the sights of a prehistoric world. We took a daylong drive to a tiny village on the outskirts of the Dja Faunal Reserve, so we'd be ready to set out first thing in the morning. But rather than renting a Land Rover to take us to our field site, the expedition would include miles and miles of backpacking and hiking.

Jen and I woke up early in the morning; put on our clothes, hiking boots, and backpacks; and set off. The packs probably weighed about fifty pounds each. We walked out of the little hut we'd stayed in and met up with our guides. For a moment, we mentally prepared ourselves for the fifteen-kilometer hike, which is over nine miles. At the end, we'd reach the field station where we'd camp for the next two weeks.

The weather was hot and humid, and the air smelled moist. It was kind of drizzling all the time. And if you looked down, you'd see the forest floor teeming with life: centipedes and giant ants and all kinds of insects.

There wasn't even a proper trail, so the guides hacked through this dense forest with machetes. It was the most intense hike I'd ever experienced, which is saying a lot. I noticed that Jen was struggling to keep up the longer we went. It reminded me of how I'd lagged behind the group during my first trip abroad to Kenya. That version of myself felt like it came from a completely different lifetime.

"Rae! Rae!" I heard Jen calling out to me from a distance. I backtracked to her.

"I'm so sorry. I don't want to cause problems, but . . . it's my feet," she said. I looked down at her hiking boots, which weren't broken in. So yes, her feet hurt. But we still had about six to seven hours left until we arrived at our destination.

"I might need to turn back. If my feet hurt this bad already, I don't think I can keep going," she told me.

I was the leader of this expedition, and my take-charge self kicked in. I sat down next to Jen.

"Jen, you've got to decide right now, because if you can't keep going, we have to go back the other way immediately," I said in a firm but gentle voice.

She started to cry, but through her tears, she bravely said, "I don't know how I'm going to do this, but we're going to keep going."

To relieve her burden a bit, I took some items from her backpack and transferred them to mine. And we kept going.

Six hours later, we finally made it to camp.

I called to Jen because I didn't see her yet.

"You're almost here, Jen! You can make it!" I wanted to instill some hope in her that the end of her torment was near.

When she came into view, her face was mottled with tears, and she was limping.

"Hey, I set up your tent for you," I told Jen. I'd arrived at the camp about forty minutes ahead of her and had rushed to get our accommodations sorted before I collapsed from exhaustion. "Why don't you go over there and take off your boots and rest for a bit?"

She limped over to her tent and removed her boots and socks, revealing bloody feet covered with blisters. After letting her feet recover for a short while, she started bandaging her wounds.

Watching her, I couldn't help but be impressed. It would've been easy for her to turn back when her feet first started hurting, but she chose to power through. And though I never would've thought it before that moment, I realized she was a good partner for the expedition. I was genuinely glad she was with me.

We had an idea that lowland gorillas existed in this forest, but we needed proof—visual observations like photos and data of where we found them and how many of them there were. Then I could report back to Johns Hopkins that this was a viable area for a gorilla study, both for the field course and for the long-term research project I'd be leading. Luckily, our two guides were from the area and assured us that yes, they'd seen these gorillas. Now we just needed to find them.

A few days later, after Jen's feet had started to heal, one of our guides, Romeo, took us on an extraordinary hike deep into the jungle to begin our search. Every so often he'd stop us and point at something: a beautiful, vibrant toucan perched in a tree. Colobus monkeys swinging through the trees, calling to one another—perhaps announcing the human interlopers. The amount of biodiversity was overwhelming.

At the same time, it seemed like we saw every kind of animal except the very one we were there to find. We didn't even see signs of gorillas, like prints or a bed made of branches and grasses.

By the fourth day, we still hadn't seen any signs. I experienced a moment of panic. I had told Johns Hopkins University that I could establish a field site in the Congo Basin where lowland gorillas lived. I didn't know what I would do if I came up empty-handed.

During our long hikes through the rainforest, my mind would wander. I started checking in with myself about whether this was the right place for me to build a new project. I had every reason to think it was. But something was nagging me.

About a week after we'd arrived, we were once again walking and wandering through the forest. I was gazing up at the beautiful, tall trees, and the guide abruptly stopped.

"Look," he said as he pointed at the ground. "Look—poop."

As a bear biologist, I love poop. I look at poop all the time because it tells us so many things. For starters, it tells me that the species was present in that location. I could take a GPS point to say this animal had been here recently. I could also dissect the poop and figure out what the animal had eaten. Poop isn't flashy or exciting, but it provides important data.

I looked down and saw the poop, but I didn't necessarily know what I was looking at.

"It's gorilla poop," the guide said.

The poop was fresh and warm. The gorilla might have been close to where we stood. But like all the previous days—and for the rest of the expedition—we never saw a gorilla with our own eyes.

In many ways, it was humorous that we came halfway around the world looking for evidence of these gorillas, and all we ended up finding was some poop. Even though it was just poop, it was proof that the gorillas lived in that rainforest. Jen and I were able to return home feeling optimistic.

When we got back, I wrote a lengthy report about the expedition and presented it to the department chair and the dean of the school. They were really impressed and said that they'd love to formally give me the opportunity to teach the field course.

I was thrilled. This was why I'd gone on the expedition, and them seeing the value in the work was validating.

I didn't know why, but I started thinking back to when Jen and I had first arrived at the field site and I'd watched Jen bandage her bloody feet. I was witnessing this woman really pushing herself beyond what she thought she was capable of. Being challenged can be rewarding, and that was true for Jen on this expedition.

But when I reflected on the ways I was challenging myself, I realized that it was as if I'd had bloody feet my whole life. I wanted to explore a new place and study a new animal, take on a new project, and do it all as a single mom. I was proud of myself, yet I was also experiencing a lot of anxiety and stress.

I considered this opportunity in the Congo Basin that I'd worked so hard for. I knew I could do amazing work there. But it didn't feel worth it. Did my whole life need to be a challenge? What if I gave myself permission to scale back? What if I allowed myself to live a life that felt easy, at least for a little while?

Surprising even myself, I went back to the department chair and explained that much had changed in my personal life, and I didn't feel that leading the field-study course and research project was the right choice for me. However, I did have a solution for moving forward with both the course and the research.

One of the guides, Romeo, was interested in wildlife ecology and had been studying it. He had taught us the most about that area while we were there. Everything these students would be learning about the forest, they'd be learning from him. So I recommended that he teach the field course. The folks at Johns Hopkins agreed, and a group of twelve students prepared to study in the Congo Basin with Romeo.

Per the university's rules, they still needed a full-time faculty member to accompany the students. Believe it or not, Jen once again volunteered to go with the students, and the whole endeavor was a huge success.

I never regretted saying no because it seemed less like I was saying no to the Congo Basin opportunity and more like I was saying yes to myself and what made sense for my life. Saying no gave me something I didn't realize I needed: permission to do less.

And once I let go of the idea that my whole life needed to be a struggle, that I had to be working hard to prove that I deserved success, that's when success started coming to me more readily.

When I was transitioning back to fieldwork, I'd felt desperate to cling to every opportunity that came my way because they were few and far between. But in those few years after divorcing Oba and severing ties with Logan, I had learned not to act out of fear and scarcity. The years on H Street were hard in many ways, but one valuable treasure I gained was faith that saying no to something didn't mean that other opportunities wouldn't appear. Instead, it allowed me to be free and available for the ones that were truly important.

Chapter 12

In Minnesota, the first snow often falls in November, blanketing the landscape of firs and lakes with a pure-white coat. The evergreen forests of the northern frontier grow silent, as the animals nestle in for their season-long lethargy, preparing dens and storing food. The winter calm envelops the forest, and its postcard-pristine magnificence will fill even the most traveled ecologists with awe.

The animals' hibernation was what brought me to the woods of Minnesota for the first time, during the latter half of its nearly eight-month-long snow season in 2020. At the start of the year, I'd been approached to host a segment of a nature show in partnership with National Geographic. I immediately accepted.

This life-changing offer for a televised segment came five years after I'd finished my PhD program and fifteen years after I'd discovered my love for wildlife ecology. This span of years between study and work showed me that it's possible to redefine what success means. A much younger Rae would have believed that success was limited to hosting my own nature series. But over time, I'd grown glad and grateful for every opportunity along the way. Success is not an arrival—it unfolds over time.

The gravity of this moment didn't have to be significant for anyone else in order for it to be significant to me. Even if only two people in the entire world watched the show, it wouldn't matter. What mattered was that I'd put a vision out to the universe in 2003. I discovered wildlife ecology and started to pursue the career in 2005. And now, in 2020, I was cast on a show, just like I wanted to be.

I made it.

It felt good to be proud of myself. It felt good to realize that sometimes dreams manifest differently from the way you had pictured—the steps can be smaller, the path can be different.

Shooting for the program would begin in Minnesota the first week of March 2020. Around this time, the United States was beginning to grasp the seriousness of the COVID-19 pandemic. As our flight west was boarding, I checked my email one last time before switching off my phone. My inbox was empty. When we landed nearly three hours later and I checked my email again, I was inundated with messages from organizations and partners saying that all business and travel were being halted. I was advised to return home as soon as I could. Confused and scared, I turned to the show's director for advice and consolation. The director decided that since we were already on location, we'd shorten the shoot from four days to two, so we could return home and shelter in place sooner.

I was also concerned for a reason many didn't yet know: I was pregnant.

After successfully building a committed long-distance relationship, Dave and I had been preparing to live together, and we set our intentions on forever. Years of therapy and self-work had prepared me to accept Dave's love and, in return, offer healthy, dependable, committed partnership. He saw parts of me that I was too afraid to look at. He held my bruised and broken bones, giving me the space and time to heal from the turbulent few years I'd had before he and I had begun thinking of each other romantically. And I was able to show myself and him what an amazing partner I could be when I was loved properly on the inside and out. The pain of healing had turned into the pleasure of fun. Less than two years into our relationship, Dave and I decided that we wanted to grow our family, quickly pivoting from a slow ramp-up into our partnership to a serious relationship decision. Months later, our wishes and prayers were answered, around the same time my professional life was hitting its stride.

The pregnancy was early on, and I didn't want anyone to treat me differently or suggest that I shouldn't participate in the filming because of its strenuous nature. However, my nausea was extreme and making me weak, and I knew in my mind that I was taking a big health risk by going forward with the opportunity. This particular segment required hiking in snowshoes, five to ten miles a day in many feet of snow in below-freezing temperatures, without access to quality food or drink. Had people known about the pregnancy, I might have been advised to bow out, to slow down.

On the day of our expedition, it was freezing cold and snowing, adding to the many inches already accumulated on the ground. The landscape was stunning. We passed numerous lakes that were completely frozen over, like precious gemstones tucked into the ground. It was such a peaceful walk to the site, the only sound our snowshoes crunching on the newly fallen flakes.

The mother bear we were tracking had a GPS collar around her neck, so we knew her approximate location. But because we were in a very snowy part of the country in a huge forest, we didn't know precisely where to find her. Luckily for us, a couple of weeks beforehand, the lead biologists, Dan and James, had hiked to the location and looked for the den. After they found it, they stuck a stick in the ground to mark the spot.

When we reached the area where the GPS led us, Dan and James found the stick with ease. They motioned for both the cameraman and me to come over and peek into the den. I walked over as quietly as I could to peer inside. At first, the den entrance looked like just a hole in the snow, no more than two feet wide and a foot tall, with branches crisscrossing the opening. Then Dan pulled out a mini flashlight and shone it into the cavity, which allowed me to see inside.

After my eyes adjusted to the near-black darkness of the den, I eventually was able to see not only movement but also the mama bear's face. She was so close we could have reached in to touch her. It was beautiful but nerve-racking because she was awake, with eyes open. She was moving slowly, not in the typical hibernation sleep, because she had growing cubs in the den with her. It's typical for mothers with nursing cubs to not sleep during winter hibernation, but to exist in a mellow, low-metabolic state. Still, we didn't want to startle her and jar her from her restful posture.

The main concern in a situation like this is that you don't know how the mother bear will react to your presence. You don't know if she'll fully wake up and attack. If bears feel threatened, even if they're out of it because they've been hibernating all winter, they can still become aggressive. Deadly, even.

Dan and James poked her gently on the shoulder with a jab stick—a sedative-filled syringe fastened to the end of a long pole—and scurried backward. We all fell silent and still, waiting for Dan and James to signal that the tranquilizer had worked and she'd fallen back asleep.

With their reassurance, we widened the hole. Dan and James stepped out of the shot, and the cameras pointed at me from different angles. I began talking, digging, digging, talking, widening the hole, and soon I heard the babies' cute yips and laid my eyes on the one closest to the entrance.

This is the best part of my job. There is nothing better in my entire career than when I do den work in the winter and get to stuff newborn bears into my jacket. It is the light of my life. My job is to keep them warm. Although they're furry bears that have been hibernating for months, they're still cubs and too tiny to thermoregulate.

The cubs were super young, but they looked healthy.

The whole filming session probably took five minutes. Once we wrapped, I turned back to Dan and James, who were checking on whether the cubs were ready to be put back with their mom. They looked preoccupied at the mouth of the den. Sensing a need for haste, I immediately started walking over with the cubs. As I drew near, I heard the guys muttering expletives in hushed voices.

"Hey, are you ready for the cubs?" I asked.

"Get back, get back, get back," they told me.

I did as instructed, unsure of what they were doing. Both men reached into the den, grabbed the mama bear by the fur, and hauled her out. At this point, I wasn't alarmed because I thought that maybe they wanted to conduct a full-body checkup on her. Maybe they'd noticed something that needed closer care. But as soon as they hauled her out, they rolled her on her back.

"She's dead," said Dan. His voice seemed impossibly loud in the stillness of the snow-covered forest.

It felt like my stomach dropped out of my body, through the layers of the earth and deep into the pits of hell.

She's dead.

I ran back over, even though they'd told me to stand back. James knelt next to the bear. "We think she suffocated," he said.

He explained that when she was tranquilized, she must have fallen asleep flat on her face. In a den that is mostly dirt and snow, there's also a lot of debris—bears will often bring insulating material into the den. James believed that the mother fell asleep on her snout and inhaled so much debris that she asphyxiated and died.

Losing a bear due to scientific or human error, or even random chance, isn't just soul-crushing because a life is lost. Such an event can also be career-ruining.

Our whole team had lost a study animal, and these cubs had lost a mother.

I returned to where the rest of the team waited. They could tell something terrible had happened, but they didn't know what. As soon as I relayed the situation, everyone had a reaction similar to my own—except two volunteers.

Instead of shaking their heads in dismay, they threw down their stuff, pushed past me, and ran to Dan and James. I instinctively followed and saw them explain a plan to Dan and James, and within seconds, all of them ran back over to the mama bear, who was still on her back. The volunteers took her arms and legs and spread them out. They then began performing CPR.

As I held her six-week-old bear cubs in my jacket, Dan, James, and the volunteers started pushing on the chest of a giant wild animal, trying to revive her. They pushed and pushed, laboring over her with laser focus.

One of the volunteers cupped his hands around the bear's snout and mouth. He started alternating between doing a few breaths and a few pumps to the chest. But both tasks were too much. He told the other volunteer, a woman about my age, that they needed to switch roles—she'd do the breath work while he pumped.

I felt the futility of the situation. Did these people know what they were doing? This wasn't a human. This was a wild bear. And she was dead. She'd been dead for at least five minutes.

All of a sudden, I heard one of the volunteers yell, "She has a pulse!"

She has a pulse!

I went back to the production team, still a few yards away, and relayed the news. But soon after, the volunteers realized that she wasn't breathing on her own. The CPR had to continue.

This went on for fifteen minutes. Pumping, breathing, pumping, breathing. The biggest men in our group all took turns pressing on her chest. Different people tried to breathe for the bear. And while she kept a pulse, she wasn't taking her own breaths. I stood nearby, clutching the cubs in my jacket and considering all possible outcomes.

At some point, everyone agreed that we should stop and figure out what to do. As one of the volunteers felt for her pulse one last time, her chest rose on its own.

"Oh, my gosh, is that her? Is she breathing?" one of the volunteers wondered aloud.

The team went right back to it. More CPR on her chest, more breathing for her. But this time, they didn't have to do it for long. Instead, the bear took over. She was breathing again. Strong breaths, all on her own.

Never in my life have I felt such relief. I had witnessed a bear come back to life.

Dan started crying. Dan, this epic bear biologist whom I'd cited in my dissertation work years beforehand, the most experienced person on our team, was an emotional wreck. I, too, shed tears of joy.

Once the mama bear started breathing on her own and her pulse returned to a normal range, the two volunteers, Dan, James, and I all grabbed a different part of her body and lowered her back into the den. Dan lay down on his belly, getting dirty and wet in the snow, to stick his head halfway into the den and make sure the mama bear was positioned with her head away from all of the debris. Slowly, I took out one cub at a time and placed them directly on her belly, so they could start nursing. We then closed up the hole as much as we could.

And we hustled out of there.

Walking back through the snowdrifts, I was struck by the haste and confidence with which the volunteers moved to save the dying bear, yet also by the fact that she'd suffocated in her den, a comfortable, familiar place. A home she'd built for herself and her cubs.

For so long, I'd struggled with the dissonance of my existence. I walked the line between care for animals and care for humans, care for career and care for family, care for partners and care for self. Like the mama bear, I was debilitated. I had stuffed my den with things I thought would bring me comfort, only to one day suffocate from their effects. Too often I found myself sprawled on the living room floor of apartments in Harlem, cabins in Tahoe, tents in Kenya, and condos in DC, lost and confused, my worth defined by people and factions that had no right to do so. Yet there I was, smothered.

But in those moments of desperation, I turned to my work—to the locales, people, and animals in which I found solace. This focused my attention externally, drawing me out of myself. Each of the ecosystems I inhabited reminded me that we all exist in webs of interdependence, navigating the dissonance that is beautifully inherent to the natural world. None of us are ever truly alone.

As an ecologist, it's my job to preserve and protect. I work to manage the relationships between animals and humans—relationships as fickle and shifting as the weather. Yet it was only when I turned my attention to my own relationships that I was fully able to appreciate my place in this world.

That shift of perspective has catalyzed the most important and self-actualized work that I have created over the course of my career. Because of it, my science is better, my media is better, my mental health is better, my energy and my parenting are better, and my communication is better.

And when I'm better, good things come.

Bears are ecosystem engineers. They spread plant and berry seeds and marine-derived nitrogen around the forest and the surrounding streams. They help maintain healthy populations of deer and other prey through predation. They break up downed logs, enabling the process of decay and returning nutrients to the soil. They keep the forest fertile and raise cubs to do the same. They provide for themselves, their offspring, and their forest peers.

As I walked through the Minnesota snow, wet and exhausted from the day's events, I smiled under my scarf. It had taken years, but I was finally able to navigate the world knowing that, just like bears, my place in this great big ecosystem was complex and diverse and necessary to its survival.

Breathing all on my own, I saw the forest through clear eyes for the first time.

Chapter 13

In August 2020, Dave and I relocated our family to Santa Barbara, California. The move occurred at a challenging time, in the middle of the first pandemic summer and at the end of what had been a difficult pregnancy. My National Geographic fellowship was ending that year, so I'd been investing a lot of time and energy in networking, pitching, and exploring career options far and wide. But I was only interested in positions that supported the intersectional nature of the work I wanted to do: science, science communication, mentorship, and media.

As the pandemic spring had rolled on, opportunities had become fewer and fewer, though there was one relationship that I'd been nurturing. I was offered the opportunity to research large carnivores, namely mountain lions and bears with The Nature Conservancy on the California Central Coast. Along with the research project came a chance to establish a home in academia, at the Bren School of Environmental Science and Management at UC Santa Barbara.

From the bed of our Washington, DC, apartment where Dave had joined me and Zuri for COVID lockdown, we scoured web pages for homes to rent and elementary schools to enroll in. Although I was given the option to continue living on the East Coast and then travel out west to conduct fieldwork every few weeks, Dave and I easily agreed that a move to California was an adventure we were interested in taking together.

With this decision came a host of complexities: Zuri's dad, Oba, would be significantly farther away. Dave's aging mother— as well as both of my parents, who still lived on the East Coast— would be thousands of miles from us and their granddaughters, and in the early days of the pandemic, it was uncertain how or when we'd be able to reunite with anyone. We didn't know a single person who lived in Santa Barbara or the surrounding region, and the cost of living was exorbitant. But we were confident in our belief that this next move toward something new was the right decision.

By December 2020, the world had been locked down for several months. Though I'd promised myself I wouldn't compulsively check my emails, I often found myself unlocking my phone and scrolling through the new messages.

I was on maternity leave after giving birth to my second daughter, Zoey, in October, and struggling with the feeling that my professional life was at a place of tremendous uncertainty. A combination of major factors was causing me to feel like a mess: postpartum depression, the extreme lack of sleep that comes with having a newborn, the physical and emotional roller coaster of attempting to breastfeed with a low milk supply, the demands of having then five-year-old Zuri on winter break after only having a mere taste of in-person school before the pandemic hit, and the uphill climb of digging ourselves out of an unstable and slightly alarming financial position after more than six months without a paycheck. Like so many other Americans, I was relying on the monthly stimulus checks and loan-payment pauses to keep us afloat.

To establish myself earlier in my career and gain respect as a bona fide scientist, I'd had to constantly conduct research. But at this juncture, my notable social media audience, reputation as a skilled public speaker, and heightened media presence ensured that I was recognized as an important voice in the scientific community. Which, after years of feeling like I had to prove myself, was a welcome relief. And because of the COVID pandemic, this became more important than ever.

The work I'd done to place myself here as both a researcher and creative voice had transitioned me into an exciting new chapter in my career. But the combination of COVID pandemic uncertainty, isolation in our new home, and postpartum depression was getting the best of me. I was beginning to forget how far I'd come and how unique I was. Traveling to distant lands to study wild animals wasn't possible, and it was unclear when the world would open up again. The global catastrophe had taken away so much from so many. For me, in addition to the fear, grief, and loss I experienced with the rest of the world, it stripped away multiple parts of my identity.

All-too-familiar questions clouded my mind with doubt. Who was I if I wasn't outdoors camping, hiking, tracking wildlife? Who was I if I was homebound while dodging a virus, nursing a newborn baby, and trying to muster the energy to entertain a five-year-old who hadn't had social interaction with peers in nearly a year? Who was I if my relationship with my soulmate and the father of my second child had drastically changed from a partnership of sexual energy and exhilarating adventure to a partnership focused on financial survival?

Avoiding the flood of spam and promotional discounts in my personal email account, I switched to my professional account and froze. At the top of my inbox was an email from the Biden administration's transition team, a group I'd had the honor of chatting with for several weeks about the importance of building a science-first, justice-centered administration. The email

was from a high-level individual whom I hadn't had a conversation with yet, and it was a brief message that read, "We'd like to
discuss an opportunity with you on the phone at your earliest
convenience." Without any idea what they could be hinting at, I
walked into the living room and signed to Dave that I was taking
a walk and a phone call. Zoey was napping on his chest, and Zuri
was watching a video on her iPad—a typical afternoon during
our pandemic winter.

"Hello, this is Rae Wynn-Grant," I offered in the most professional voice I could muster when my call went through. I
exchanged pleasantries with the woman on the other end of the
line, who introduced herself as a leader in the Biden administration, tasked with selecting the highest-quality people to fill his
cabinet.

"We'd like to offer you a presidential appointment," she stated.

I stopped in my tracks, heart beginning to pound. I wasn't
entirely sure what that meant, but I knew it was major.

"We'd like to position you to be the principal deputy director of the US Fish and Wildlife Service. After congressional
approval, you would become the director of the US Fish and
Wildlife Service. We think you would be fantastic in this role,"
she continued.

"Oh my gosh," I managed to squeak out.

Originally, I'd been tapped to help the administration staff the
Department of the Interior by recommending people for various
positions, providing guidance on what skills and backgrounds
would be most useful. But when things shifted, and they began
interviewing me for high-level positions, I never thought they'd
offer *me* a job, let alone a position so prominent and visible that
I'd be one of the leaders in government. The US Fish and Wildlife
Service is the country's federal-level conservation group, working

nationally and internationally to study and protect wild animals. I'd admired the group for years, explored their hiring pages countless times, and dreamed of being recognized by them for my work in wildlife conservation. Now, here I was with an offer on the table to lead the entire service for the next four years, and possibly beyond. An active scientist had never held the role before and, of course, the optics of a millennial Black woman like me taking the reins of the organization were profound.

This decision couldn't be made overnight, but, unfortunately, my days to give my answer were limited.

On Zuri's first day back to in-person kindergarten in January, I piled Dave and baby Zoey into our car and set out on a drive. I had to decide whether I'd accept the offer to lead conservation for the Biden administration. Amid the whirlwind of virtual school, a sleepless newborn, and an interrupted maternity leave, I hadn't found the space to think things through with myself, let alone discuss it with my partner. After receiving the email, I'd texted both of my parents and my brother to give them the gist, and they, too, were waiting with bated breath to hear my choice.

I was originally supposed to make my decision by Friday, January 8, but the Capitol riots on Wednesday, January 6, created a well-understood delay in all of the work around transitioning administrations.

As the scenes unfolded that Wednesday, news poured in from my DC-based family, friends, and colleagues. At one point I texted Carmen, the concierge at the apartment building near Capitol Hill that we had moved out of only five months prior. I called my father and stepmother to make sure they weren't planning to go out as counterprotesters—which they often did but had cut back on because of COVID risks. Updates came through several group-text chains from close friends who worked in the

White House and were locked down inside their offices, worrying that they might be attacked. One of my friends, a Black woman my age who worked for one of the handful of Black female Congress members, described her whole office sheltering under their desks and crying, nearly certain that because of their identities, they'd be targeted first.

The Capitol riots added another complex layer to my decision. Only a couple of years prior, I had seen DC as a potential long-term home for myself and my family. But the events in the months leading up to this potentially life-altering choice were plagued with concerns and questions about my position and complacence in a system built against me and my community. After all, the Fish and Wildlife Service had a law enforcement branch—a by-product of the job that I, at that point in my social justice consciousness, didn't feel comfortable with.

I called my contact at the Biden administration and asked what flexibility I'd have to be able to continue my science research projects, working on television and book projects, and using my social media for informal science communications. Her response was gentle but firm: "With a position like this, it would need to be your only professional work, and there might be room for some media work, as long as it was released after your term in the position has ended."

This didn't surprise me, nor did I disagree with the requirement. But I could feel my heart screaming at me to stay the course that I'd fought and sacrificed for. The life I'd built for myself was one worth protecting. And the career I'd curated was one worth pursuing. Undertaking this massive career opportunity could disrupt much of the life I'd worked so hard to build. And I couldn't let that happen.

Sitting in the driver's seat that January day, I turned to look at Dave. He smiled, and I glanced into the rearview mirror to smile at Zoey playing with her feet. As the turmoil of the outside world violently churned around us, I knew I was content. My focus returned to the road. I knew what my reply would be.

"ARE THOSE SEALS or sea lions?" asked the graduate student I'd brought to my field site. She excitedly peered through her binoculars, spotting more marine mammals than she'd ever seen in one place. "Dr. Wynn-Grant, what do you think? I'd guess seals because they have all of those spots, and it's spotted seals that live around here, right?

"Dr. Wynn-Grant?" she said, trying again to get my attention and elicit a response.

I was standing right next to her but I couldn't speak. My gaze had been on the herd of seals swimming close to shore, their dark heads bobbing in and out of the water. Then something bigger captured my attention: a dark shadow making larger waves just beyond the seals. By the time my brain registered what was happening, the creature's back curved across the surface of the water, shooting water vapor far into the air from its blowhole.

I always said I'd cry when I saw my first whale. And I did. Tears streamed down my face as I watched a gray whale's dorsal hump and thick tail emerge from the water, as the animal then dove, creating a giant splash. The whale was so large and so close to the shore that I could see it clearly without binoculars. It was solo and swimming northward, close to where the southern headwaters meet the northern headwaters of the Pacific Ocean.

I was in the field in the spring of 2023, after the White House job offer was a distant memory, after vaccines and the cycles of

COVID variants were the new norm, and after Dave, Zuri, Zoey, and I had fully rooted ourselves in California.

The students I was with on the Jack and Laura Dangermond Preserve that day would help me with my first big push of field-work, to investigate the unique ecologies of the mountain lions that might use that area, with a special focus on why our camera traps kept capturing images of them walking toward the beach. At this mid-career stage, I felt unsure of where my time and focus would take me during the next decade, but instead of anxiety, I felt secure in the fact that no matter what, it would be my choice. I had rooted myself well enough in several areas of expertise that I knew my career moves would no longer be made out of despera-tion or from a mindset of lack, but from the heart.

Dave continued to prove himself to be the dependable part-ner I needed, managing our lives in a way that allowed me to spend time in the field and travel for public speaking and media appearances.

Years ago, when I'd left the American Museum of Natural History on a leap of faith, I told myself and the universe that I wanted my work to revolve around talking to the public about nature, mentoring people who didn't yet know they wanted to be scientists. And I wanted that work to simultaneously feel like freedom, for myself and my family.

We often believe that success looks like money or status or material possessions, and if I'd stayed in that mindset, I'd be miss-ing out on a lot. I don't own a home or car; I don't have a savings account for myself or my children. But I believe those things can be acquired over time, and I'm making plans for the future.

Through my work in the field, and the observation, persever-ance, and determination it requires, I've learned that it can take years, decades, or a lifetime to see all the things you want for your life to line up—and you might have to fight to achieve even the tiniest part of your goal. However, unless you're keeping your eyes on the horizon, sometimes it can be hard to notice when what you've been working toward has entered your line of vision.

DAVE AND I still live in California, where the story of my life began so long ago. As I reflect on the time spent in between then and now—the places I've visited and the beings for which I've cared—I feel an overwhelming sense of joy, accomplishment, and pride.

Oba and I talk, plan, and co-parent almost every day. He still lives in New York City and has established a rhythm to his California visits, and Zuri is now old enough to travel with him back to the East Coast on her breaks. He is a great dad, and we're making it work.

I have sacrificed, I have stumbled, and I have fallen flat. I have steadied myself, bet on myself, and entered into the unknown countless times. I have been a formal student of nature for twenty years and counting. And now I have options—I can put the work down, bow out, hand it off. I can do something more traditional with my time, with my motherhood, with my life. But I won't. I'm not done.

After all, the wild is still calling.

Epilogue

In many ways, I'm still that nineteen-year-old taking a chance on a study-abroad program, forcing herself out of a comfort zone and diving headfirst into what I think is the right path for me. These days, my footing in the wild is more stable, but the path I find myself forging today isn't through the wilderness that I've grown to know and love as my second home. I'm taking steps into a whole new world—a far trickier jungle, where many of the survival rules I've learned in nature don't apply.

I had long believed there was one way to be a real scientist: you participated in science research day in and day out, published peer-reviewed articles in academic journals, attended conferences to learn about (and critique) new work, and led a lab of graduate students at a research-focused university. Although I'd never imagined that strictly academic scientist life for myself, I couldn't shake the idea that hosting a television show would discredit any science I had ever done. It would turn me from Dr. Rae Wynn-Grant, BS, MS, MA, PhD, a leader in ecology and evolutionary biology who'd graduated from some of the world's most rigorous programs, into just another image-conscious starlet seeking fame.

Months, years, and decades thinking about this lifelong goal had passed by, and I was coming to terms with the fact that it might not be in the cards. Then, when I least expected it, I got the call.

I remember every detail about where I was standing in my office, the slightly chilly temperature, how hungry I was after working through lunch. I looked at my phone and saw the name on the screen. It was Jennifer Wulf, the Vice President of Brand Marketing at Mutual of Omaha, who had no reason to call me out of the blue unless she had something important to say.

"Would you be interested in cohosting *Wild Kingdom*?" Jen asked. I could envision it: being in front of a camera, hiking through a forest or savanna, describing the wildlife and the land-scape to the viewers at home, like the nature-show hosts of my youth had done so well.

I sat down and took a deep breath.

"Jen, are you kidding? Absolutely!"

Mutual of Omaha's *Wild Kingdom* has a long and important legacy in American media. Airing first on NBC in 1963 with hosts Marlin Perkins and Jim Fowler, every Sunday night it gave audiences across the country a peek into the world of wild animals in wild places. At a time when television didn't offer nearly as many shows as today, *Wild Kingdom* was a staple in many, if not most, households. In 1985, the year I was born, Peter Gros took over as host on the show, traveling around the world and bringing the wilderness to living rooms while continuing the show's legacy. As a child in California, I'd tune in to *Wild Kingdom*, and I enjoyed the reruns with the original hosts.

This was one of the shows that had inspired my entire career and, most importantly, set me on the path to becoming a scientist. And now here I was, on the phone with a Mutual of Omaha executive asking me to be the newest host, helping to bring the show back to NBC and American audiences once again, sixty years after it debuted.

A swell of happiness building in my body, I smiled as I told her, "It would be a dream come true."

Acknowledgments

THE WRITING OF THIS BOOK did not come easily, quickly, or smoothly. In fact, the entire concept of capturing my origin story was brought to me by my literary agents, Jamie Chambliss and Steve Troha. Thank you for contacting me out of the blue with this idea and for continuing to push me forward when the initial lift of writing the proposal seemed too heavy for me to carry. Thank you to John Jardin and Eloise Davenport, far beyond their years, who shaped my proposal as interns at Folio.

Thank you to Mike Jackson, John Legend, and Ty Stiklorius at Get Lifted Film Company, for seeing me and my potential as an author, thought leader, and inspirational character through the pages of the book proposal. Thank you for believing in this book and my stories from the idea stage to the final product. Thank you for recognizing the creativity hidden within my factual science exterior.

To Marcie Cleary, you are far more than a brilliant and trustworthy attorney. You are a dear friend with a crystal ball, who could see my potential, success, and worth far before the evidence was there. From graduate school to first-time motherhood to finally having a business relationship together, the Black Girl Magic you have instilled within me for years is sacred, and I am beyond grateful for your guidance and loyalty.

Thank you to Molly Stern for breaking glass, disrupting the industry with Zando, and offering me the opportunity to be one of a few select authors worthy of working with your company; Quynh Do, my original editor, who welcomed me into Zando with open arms and positivity; and Sarah Ried: there isn't enough space on these pages to detail what your partnership—editorial, intellectual, and moral—has meant to me in this process. Thank you for bringing me to the offices at Zando, for treating me like a star even when I was far behind in my progress, and for making promises and bending rules in order to make the impossible possible.

To Amanda C. Bauch, you spent the summer, fall, winter, spring, and another summer by my side as I started, stopped, panicked, gave up, picked myself back up, reengaged, overcommitted, apologized, confessed, surrendered, and showed back up. Thank you for supporting me through the memoir process, offering your empathy, modeling best practices, and sacrificing your own blood, sweat, and probably tears to get these words on the page.

Thank you to Rachel Aronoff and Caroline Hadilaksono, for the countless hours you spent coaching me through storytelling and helping me recall the most intricate details from memories both painful and pleasant.

To Lauren Andrews and Yessica Rocha, you two took exceptional care of my children during my late nights at the office and my far-off fieldwork. This book, and my career, could not have been accomplished without you and the other caregivers who offered love, affection, and deep caring to my girls in my absence.

To my beloved Elizabeth Hinton, you gave me a deep and personal example of how a busy, overcommitted millennial working mother and academic could create a powerful book. You showed me it is worth it, and you were right.

To my family—the blood family, the in-laws, the step-daughters/sisters/mothers, the chosen family, the friends who have become family, to all of the Wynns, the Grants, the Wynn-Grants, the Sutlers, the Davises, the Seligmans, and the Bennetts, thank you for this life, the rich history of us, the love, the care, the support, and the future.

Dave, thank you for believing I could write this book. Thank you for making space for me and for shouldering so many family obligations and logistics and so much emotional energy so that I could create. Thank you for loving me so deeply, for the grace, the faith, and all of the fun.

To my children—it's because of both of you that I am well. Your lives are a miracle and a gift, both to me and to the world. May some of the stories and messages in this book help to guide you through your own life's adventures and serve as a reminder that your North Star is yours and yours alone.

About the Author

DR. RAE WYNN-GRANT is a wildlife ecologist who studies the impact of human activity on carnivore behavior and ecology. She received her BS in environmental studies from Emory University, her MS in environmental studies from Yale University, and her PhD in ecology and evolution from Columbia University. She is the cohost of *Mutual of Omaha's Wild Kingdom Protecting the Wild* on NBC and hosts the award-winning podcast *Going Wild with Dr. Rae Wynn-Grant*, produced by PBS. She is a National Geographic Explorer and has worked with National Geographic on a variety of televised nature programs, as well as a twenty-city speaking tour. Dr. Wynn-Grant has been featured in *Vogue*, *Forbes*, the *New York Times*, and the *Los Angeles Times*, among many others. She lives in California with her family.